T0221392

# Soil Remediation

## Applications and New Technologies

# Soil Remediation

## *Applications and New Technologies*

*Editors*

**José Tomás Albergaria**

Requimte, Instituto Superior de Engenharia do Porto
Porto
Portugal

**Henri P.A. Nouws**

Requimte, Instituto Superior de Engenharia do Porto
Porto
Portugal

CRC Press
Taylor & Francis Group
Boca Raton  London  New York

CRC Press is an imprint of the
Taylor & Francis Group, an **informa** business

A SCIENCE PUBLISHERS BOOK

Cover illustrations reproduced by kind courtesy of Arcadis Nederland B.V.

CRC Press
Taylor & Francis Group
6000 Broken Sound Parkway NW, Suite 300
Boca Raton, FL 33487-2742

First issued in paperback 2020

© 2016 by Taylor & Francis Group, LLC
CRC Press is an imprint of Taylor & Francis Group, an Informa business

No claim to original U.S. Government works

ISBN-13: 978-1-4987-4361-7 (hbk)
ISBN-13: 978-0-367-78325-9 (pbk)

This book contains information obtained from authentic and highly regarded sources. Reasonable efforts have been made to publish reliable data and information, but the author and publisher cannot assume responsibility for the validity of all materials or the consequences of their use. The authors and publishers have attempted to trace the copyright holders of all material reproduced in this publication and apologize to copyright holders if permission to publish in this form has not been obtained. If any copyright material has not been acknowledged please write and let us know so we may rectify in any future reprint.

Except as permitted under U.S. Copyright Law, no part of this book may be reprinted, reproduced, transmitted, or utilized in any form by any electronic, mechanical, or other means, now known or hereafter invented, including photocopying, microfilming, and recording, or in any information storage or retrieval system, without written permission from the publishers.

For permission to photocopy or use material electronically from this work, please access www.copyright.com (http://www.copyright.com/) or contact the Copyright Clearance Center, Inc. (CCC), 222 Rosewood Drive, Danvers, MA 01923, 978-750-8400. CCC is a not-for-profit organization that provides licenses and registration for a variety of users. For organizations that have been granted a photocopy license by the CCC, a separate system of payment has been arranged.

**Trademark Notice:** Product or corporate names may be trademarks or registered trademarks, and are used only for identification and explanation without intent to infringe.

---

#### Library of Congress Cataloging-in-Publication Data

---

Names: Albergaria, Jose T. V. S. de, author. | Nouws, Hendrikus P. A., author.
Title: Soil remediation : applications and new technologies / Jose T.V.S. de
Albergaria and Hendrikus P.A. Nouws.
Description: Boca Raton, FL : CRC Press, Taylor & Francis Group, 2016. |
Includes bibliographical references and index.
Identifiers: LCCN 2015041213 | ISBN 9781498743617 (hardcover : alk. paper)
Subjects: LCSH: Soil remediation.
Classification: LCC TD878 .A47 2016 | DDC 628.5/5--dc23
LC record available at http://lccn.loc.gov/2015041213

---

**Visit the Taylor & Francis Web site at**
**http://www.taylorandfrancis.com**

**and the CRC Press Web site at**
**http://www.crcpress.com**

# Preface

Since the nineteenth century, the development of societies led to the exponential growth of industrial and other economic activities that have originated thousands of cases of land- and groundwater contamination. During the last century these occurrences became more frequent and problematic because of the increase of the world's population as well as the fast development of science that promoted the use of new technologies, some of them more polluting than their antecessors. More recently, the valorization by modern societies of environmental aspects and its preservation implied the inversion of this tendency and led to the reduction of the rate of new contamination cases. Nevertheless, the contamination heritage still requires the urgent implementation of remediation actions. Another aspect that should not be ignored is the appearance of new contaminants that started to be considered only recently because of the advances in several scientific areas such as analytical methods and toxicology. These new contaminants are called Emerging Contaminants and pose a new challenge to industry, service providers, regulators and the scientific community.

A recent report from the European Environment Agency (2014) indicates that there are near 2.5 million potentially contaminated sites in Europe and that the waste disposal & treatment and the industrial & commercial activities are responsible for 72% of these sites. However, these numbers only correspond to the data of some European countries, so the real numbers should be much higher. A similar situation occurs in the United States of America; neither the U.S. Department of Agriculture nor the Department of Interior have a complete inventory.

In this light a more extensive evaluation of contaminated sites is essential and extremely urgent. Where necessary, the most appropriate remediation technology should subsequently be implemented. It is therefore useful to collect the knowledge about the most recent advances on the existent remediation technologies as well as the development of new remediation options that use innovative technologies and new materials to face new contamination challenges namely with emergent contaminants that require specific, and in several cases expensive treatments.

The objective of this book is to gather valuable information on soil remediation and related issues such as case studies, decision making tools (e.g., Life Cycle Assessment) and new and innovative remediation technologies (e.g., nanoremediation), that could aid stakeholder activities.

We are thankful to all the contributors for their collaboration that made the present book possible. We also extend our appreciation to Science Publishers for their extraordinary support.

# Contents

*Preface*                                                                                    v

1. **Electrokinetic Remediation and Hybrid Technologies for the**                            1
   **Treatment of Organic Pollutants**
   *María Ángeles Sanromán, Olalla Iglesias, Emilio Rosales* and *Marta Pazos*

2. **Bioremediation of Chlorinated Ethenes**                                                21
   *Anthony S. Danko* and *James M. Cashwell*

3. **Eco-Labelling of Petrol Stations: A Successful Experience in Brazil**                  39
   *Angelo R.O. Guerra* and *Francisco de A.O. Fontes*

4. **Biodegradation of Pyrethroid Pesticides**                                              59
   *Idalina Bragança, Valentina F. Domingues, Paulo C. Lemos* and
   *Cristina Delerue-Matos*

5. ***In Situ* Chemical Oxidation (ISCO)**                                                  75
   *Aurora Santos* and *Juana Mª Rosas*

6. **Soil Contamination and Life Cycle Assessment**                                         95
   *Florinda Figueiredo Martins*

7. **Nanoremediation with Zero-valent Iron Nanoparticles**                                 108
   *S. Machado* and *J.G. Pacheco*

8. **Managing Contaminated Groundwater—Novel Strategies and**                              121
   **Solutions Applied in The Netherlands**
   *Hans Slenders, Rachelle Verburg, Arnold Pors* and *Anouk van Maaren*

9. **Biological Techniques to Remediate Petroleum Hydrocarbons in**                        139
   **Contaminated Environments**
   *Nazaré Couto* and *F. Javier García-Frutos*

10. **Phytoremediation of Salt Affected Soils**                                            149
    *João M. Jesus, Anthony S. Danko, António Fiúza* and
    *Maria-Teresa Borges*

*Index*                                                                                    165

# Electrokinetic Remediation and Hybrid Technologies for the Treatment of Organic Pollutants

*María Ángeles Sanromán,*[1,a] *Olalla Iglesias,*[1,b] *Emilio Rosales*[1,c] and *Marta Pazos*[1,*]

## ABSTRACT

Electrokinetic remediation is an *in situ* technology that has been successfully applied for the removal of inorganic pollutants since the past 20 years. In this technique, an electric field is applied to promote the movement of contaminants toward the electrode chambers, located in the polluted soil, from where the pollutants can be extracted. However, in the remediation of soils polluted with organic compounds, the main limitation is that the pollutants should be soluble in the interstitial soil fluid to be transported by the action of the electric field. Therefore, the application to low solubility pollutants such as, for example, hydrophobic organic pollutants is limited. Nowadays some technique improvements can be applied in order to solve this problem and the application of this technique alone or combined with other processes like Fenton or bioremediation is giving satisfactory results in organic pollutant removal. This chapter describes the foremost principles to carry out the electrokinetic remediation of soils contaminated with organic pollutants, as well as the different alternatives to improve the process and new methodologies fororganic pollutants removal by using hybrid technologies.

[1] Departamento de Ingeniería Química, Grupo de Bioingeniería y Procesos Sostenibles, Universidad de Vigo, Edificio Isaac Newton, Despacho 21, Lagoas - Marcosende 36310 Vigo.
[a] Email: sanroman@uvigo.es
[b] Email: olaiaic@uvigo.es
[c] Email: emiliorv@uvigo.es
[*] Corresponding author: mcurras@uvigo.es

## Introduction

### *Fundamentals*

Electrokinetic remediation process arises as an effective technique for the *in situ* treatment of soils with organic and inorganic pollution. It is based on a flushing process generated from the action of an electric current that allows the transport of pollutants even when the soil is characterized with low permeability. This technique can be found in the literature with several terms such as electrokinetic remediation, electrokinetics, electroremediation, or electrochemical remediation, among others (Acar and Alshawabkeh 1993).

The principle of electrokinetic remediation consists of the controlled application of low intensity direct current through the soil between appropriately distributed electrodes (Fig. 1). One of the most important advantages of this technique is its efficiency for the treatment of low hydraulic permeability soils, where other techniques such as pump-treat are not adequate (Acar and Alshawabkeh 1993).

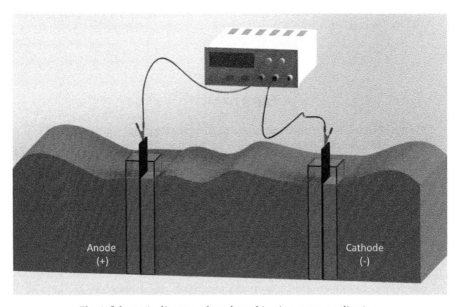

**Fig. 1.** Schematic diagram of an electrokinetic process application.

The applied electric current induces the transport of species in the interstitial soil fluid, depending on their electrical charge, towards the anode or cathode chambers (mainly due to electro-osmosis and ion migration) that is coupled with electrolysis and geochemical reactions.

### Transport Mechanisms

The application of the direct current in the soil induces the generation of fluid transport and the mobilization of species due to reactions that take place in the soil and the electrode chambers. The main electrokinetic transport mechanisms include electrophoresis, electro-osmosis, and ion migration (electromigration) (Fernandez et al. 2009). These three phenomena can be summarized as follows:

- Electrophoresis is the migration of charged colloids under an electrical potential gradient towards the electrode with opposite charge. In a saturated system such as a soil, the electrophoresis process has less importance because the solid phase is in stationary status (Yu and Neretnieks 1996).
- Electro-osmosis involves interstitial water pore transport at the solid/liquid interface. Most of the particles on the surface of the soil grains are negatively charged due to isomorphic substitutions and the presence of broken links. These negative charges attract the positive charges present in the interstitial fluid (cations) in order to reach electroneutrality. Therefore, there is an inert layer of cations on the surface of the soil pore named fixed layer. These cations do not have enough density to compensate all the negative charges; thus, a second layer of cations, which are further away from the negative groups, is not very strongly linked and constitutes a mobile layer. The combination of both layers is known as the double diffuse layer of cations. When an electrical gradient is applied to a soil-water system, the particles are fixed, but the mobile diffuse layer moves and the solution is carried with it towards the cathode. In most cases, the soil is negatively charged and the electro-osmotic transport generates a hydraulic flux towards the cathode chamber.
- The ion migration is a diffusive transport generated by electrical forces on ions. Anions are transported towards the anode (positive electrode) and cations are transported towards the cathode (negative electrode).

In addition to these mechanisms, common mass-transport mechanisms, such as diffusion or convection, and the physical and chemical interactions of the species with the medium also take place; however, their relative importance is almost negligible (Fernandez et al. 2009). Therefore, Fig. 2 shows the main phenomena that have a clear influence on the net pollutant transport in electrokinetic treatments of soils.

## Advantages and Disadvantages

Electrokinetic remediation has attracted interest among scientists and governmental officials in the last decade, due to several promising laboratory and pilot-scale studies and experiments. However, regardless of promising results, this method has its own advantages and drawbacks.

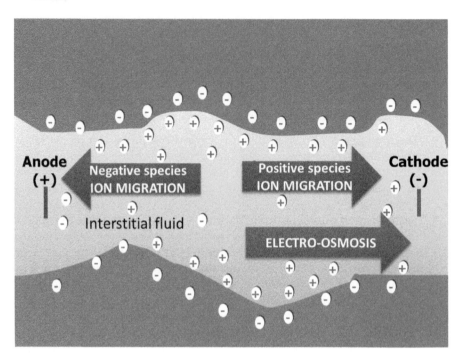

**Fig. 2.** Main transport mechanisms in the electrokinetic treatment of polluted soils.

### *Advantages*

Electrokinetic remediation has become a technology with great prospects (Huang et al. 2012; Niroumand et al. 2012). The main advantages of this process are described below:

### *Remediation of low permeability soils*

Electrokinetic remediation is the most effective treatment for low permeability soils because it can operate even at a low hydraulic gradient, when the application of traditional technologies is restricted. In clayey sediments, hydraulic flow through pores can be extremely limited. This technology has been an effective method of inducing movement of water, ions, and colloids through fine-grained sediment (Murillo-Rivera et al. 2009; Reddy et al. 2009). Whereas values of hydraulic conductivity can vary among many orders of magnitude for different soil types, the electro-osmotic conductivity lies in a narrow range of $1 \cdot 10^{-9}$ to $1 \cdot 10^{-8}$ $m^2 \cdot s^{-1} \cdot V^{-1}$. Thus, an electric field is a much more effective force driving the fluid through fine grained soils of low hydraulic conductivity than a hydraulic gradient (Niroumand et al. 2012).

## *In situ remediation*

*In situ* remediation technologies for contaminated soils include isolation and containment, soil flushing, biochemical treatment, phytoremediation, and electrokinetic remediation. The advantages of electrokinetic remediation techniques rest with possible applications under sealed surfaces and in fine-grained soils. In the *in situ* treatment of polluted soil, the main considerations are the spatial arrangement and spacing between the electrodes that significantly influences the distribution of the potential isolines (Alshawabkeh et al. 1999), the electrode chambers construction and the electrolyte which should be formulated depending on the type of soil and its pollution (Wieczorek et al. 2005).

## **Disadvantages**

### *pH*

The electrokinetic remediation process is highly dependent on pH conditions during the application, which influences the release of pollutants into the interstitial soil fluid and modifies the transport mechanisms. During the remediation process, the application of the direct current induces chemical reactions upon the electrode surfaces. The dominant and most important electron transfer reactions that occur during the electrokinetic process are the electrolysis of water:

Anode $H_2O \rightarrow 2H^+ + (1/2)O_2(g) + 2e^-$ (1)

Cathode $2H_2O + 2e^- \rightarrow 2OH^- + H_2(g)$ (2)

As it is represented in reaction (1), the protons are generated in the anode chamber and as consequence of the applied direct current an acid front is carried towards the cathode by electrical migration. Therefore, generated protons travel through the soil decreasing the pH. At the same time, hydroxyl anions are produced in the cathode chamber (2) and they travel through the soil towards the positive electrode. This fact produces an increase in the pH near the cathode. These changes in the pH of the interstitial fluid drastically affect the characteristics of the soil and the evolution of the electrokinetic remediation process.

The basic front increases the precipitation of cationic pollutants, which may inhibit their removal by the proposed technology. On the other hand, the low pH environment promotes the solubilization of inorganic compounds, facilitating their transport by electro-osmosis and electromigration. However, the low pH in the interstitial fluid of soil modifies the zeta potential (the electric potential at the junction between the fixed and mobile parts of the double layer) of soil particle surfaces, reverting the electro-osmotic flow from the cathode towards the anode (Probstein and Hicks 1993). These effects may diminish each other depending on the chemical state and polarity of charges of the inorganic compounds of soil. As a result, there are many complexities of the electrochemical processes for researchers to untangle. Moreover, due to the low pH at the anode compartment, the anode is easily corroded. This becomes a serious concern when the process scale is enlarged. Anode material should, therefore, be resistant to corrosion in order to supply stable electricity for long-term operations (Kim et al. 2011). Nonetheless, many enhancement techniques on electrokinetic remediation of soils were developed on the basis of pH control (Alcántara et al. 2008; Cang et al. 2013; Lee and Yang 2000; Saichek and Reddy 2003); although achieving these acidic conditions might be difficult when the soil buffering capacity is high (Pazos et al. 2009). One plausible solution is the inversion of polarity, a technique that consists in inverting the electric field for a short period in order to control pH in the electrode chamber (Alcántara et al. 2008; Pazos et al. 2006). In these conditions, the electrical resistance is reduced and the transport of pollutants is improved.

### Remediation of liquid collected from soil

Additionally, to obtain the total degradation of mobilized pollutants from the contaminated soil, the liquid collected by electrokinetic remediation must be treated. To overcome this problem, innovative processes that combine soil electrokinetic remediation and liquid electrochemical oxidation, for the degradation of organic compounds present in a polluted soil, were developed and evaluated (Sanromán et al. 2005; Gómez et al. 2009).

## Electrokinetic Treatment for Organic Pollutants Removal

Considering the different physical and chemical properties of the organic contaminants compared to the properties of inorganic pollutants, the operating conditions of the electrokinetic treatment should be modified in order to increase the removal efficiency. Therefore, some enhancements should be performed during the electrokinetic treatment in order to favor the remediation of organic pollutants by the electric field action (Table 1).

The simplest case is the electromigration of soluble organic pollutants, such as dyes, from the solid matrix. In this field, Pazos et al. (2007) determined that in the treatment of clay polluted with Reactive Black 5 dye, a complete removal was achieved using $K_2SO_4$ as processing fluid and pH control in the anode chamber. The used electrolyte enhanced the desorption of Reactive Black 5 from the kaolin matrix and the pH control favored the alkalization of the system and, at high pH values, Reactive Black 5 was ionized and migrated towards the anode chamber (Fig. 3 right).

There, the dye was degraded by the electro-oxidation on the anode surface. The previous process can be simplified when the electrolyte used in the electrode chambers contributes to the basic pH in the soil.

**Table 1.** Working conditions and removal efficiency of several electrokinetic remediation processes applied at different polluted soils.

### Electrokinetic remediation

| Pollutant | Conc. (mg/kg) | Voltage/Intensity | Soil | Electrolyte | Removal (%) | Reference |
|---|---|---|---|---|---|---|
| Diesel | 10,000 | 40 V | Natural soil | Sodium chloride | 55 | (Tsai et al. 2010) |
| Dinitrotoluene | 500 | 1 V/cm | Kaolin | Sodium bicarbonate, calcium chloride, magnesium chloride | 45 | (Reddy et al. 2011a) |
| Hexachlorobenzene | 100 | 1.5 V/cm | Kaolin | Sodium hydroxide | 63 | (Pham et al. 2009) |
| Lindane | 1,000 | 1 V/cm | Kaolin | Sodium bicarbonate, calcium chloride, magnesium chloride | 57 | (Reddy et al. 2011a) |
| Lissamine Green B | 300 | 30 V | Kaolin | Disodium hydrogen phosphate | 94 | (Pazos et al. 2008) |
| Pentachlorophenol | 100 | 1 V/cm | Kaolin | Sodium bicarbonate, calcium chloride, magnesium chloride | 75 | (Reddy et al. 2011a) |
| Pentachlorophenol | 100 | 2 V/cm | Kaolin | Sodium bicarbonate, calcium chloride, magnesium chloride | 78 | (Reddy et al. 2011b) |
| Phenanthrene | 1.9 | 14 V | Kaolin | Sodium chloride | 25 | (Ko et al. 2000) |
| Phenanthrene Fluoranthene | 100 | 1.5 V/cm | Kaolin | Sodium hydroxide | 84 90 | (Pham et al. 2009) |
| Reactive Black 5 | 390 | 30 V | Kaolin | Potassium sulphate | 94 | (Pazos et al. 2007) |

*Table 1. contd....*

*Table 1. contd.*

**Electrokinetic remediation**

| Pollutant | Conc. (mg/kg) | Voltage/Intensity | Soil | Electrolyte | Removal (%) | Reference |
|---|---|---|---|---|---|---|
| Indigo | 1,132 | 30 V | Sand | - | 76 | (Sanromán et al. 2005) |
| Total Petroleum Hydrocarbons | 12,500 | 2 V/cm | Oil contaminated soil | Potassium phosphate dibasic, potassium phosphate monobasic, magnesium sulfate, ammonium nitrate | 28.6 | (Dong et al. 2013) |

**Electrokinetic-Fenton**

| Pollutant | Conc. (mg/kg) | Voltage/Intensity | Soil | $H_2O_2$ concentration | Removal (%) | Reference |
|---|---|---|---|---|---|---|
| Diesel | 10,000 | 40 V | Natural soil | 8% | 97 | (Tsai et al. 2010) |
| Phenanthrene | 200 | 30 V | Kaolin | 7% | 80 | (Kim et al. 2005b) |
| Phenanthrene | 200 | 140 V | Sandy soil | 5% | 81.6 | (Park et al. 2005) |
| Phenanthrene | 200 | 30 V | Commercial soil | 7% and sodium dodecyl sulfate | 70 | (Park and Kim 2011) |
| Phenanthrene | 200 | 30 V | Hadong clay | 7% and sodium dodecyl sulfate | 75–80 | (Kim et al. 2007) |
| Phenanthrene | 500 | 3 V/cm | Kaolin | 10% | 99 | (Alcántara et al. 2008) |
| Total Petroleum Hydrocarbons | 11,680 | 3 V/cm | Marine sediments | 10% and sodium ethylenediamine tetra acetic | 90 | (Pazos et al. 2013) |
| Trichloroethylene | 233 | 1 V/cm | Loamy sand Sandy loam | 4000 mg/L | 39.5–88 | (Yang and Liu 2001) |

**Electro-bioremediation**

| Pollutant | Conc. (mg/kg) | Voltage/Intensity | Soil | Microorganism | Removal (%) | Reference |
|---|---|---|---|---|---|---|
| 2,4-Dichlorophenol | 800–1,200 | 0.89 A/m$^2$ | Silt soil | *Burkholderia* spp. RASC c2 | 87.1 | (Jackman et al. 2001) |
| Crude oil | 50,000 | 1 V/cm | Natural soil | Bacterial consortium | 45 | (Li et al. 2010) |
| Diesel | 20,000 | 1–2 V/cm | Loamy-sand soil | Indigenous microorganisms | 64 | (Pazos et al. 2012) |
| Mixture of PAHs | - | 1 V/cm | Industrial contaminated soil | Isolated bacteria | 26 | (Li et al. 2012) |
| Pentadecane | 5,000 | 0.63 mA/cm$^2$ | Kaolin | *Pseudomonas* sp. | 77.6 | (Kim et al. 2005a) |
| Phenanthrene | 200 | 20 V | Natural soil | *Bacillus subtilis* | 80 | (Xu et al. 2010) |
| Phenanthrene | 4,501 | 0.5–0.6 V/cm | Historical polluted soil | *Novosphingobium* sp. LH128 | - | (Niqui-Arroyo et al. 2006) |
| Total Petroleum Hydrocarbons | 2,777 | 0.5–0.6 V/cm | Historical polluted soil | *Novosphingobium* sp. LH128 | - | (Niqui-Arroyo et al. 2006) |
| Total Petroleum Hydrocarbons | 12,500 | 2 V/cm | Oil contaminated soil | Bacterial consortium | 72.8 | (Dong et al. 2013) |

**Electrokinetic treatment with surfactants**

| Pollutant | Conc. (mg/kg) | Voltage/Intensity | Soil | Surfactant | Removal (%) | Reference |
|---|---|---|---|---|---|---|
| Benzo[a]pyrene | 300–400 | 3 V/cm | Kaolin | Brij 35 | 76 | (Gómez et al. 2009) |
| Gasoil | 20,000 | 30 V | Natural soil | Rhamnolipid | 86.7 | (Gonzini et al. 2010) |

*Table 1. contd....*

*Table 1. contd.*

**Electrokinetic treatment with surfactants**

| Pollutant | Conc. (mg/kg) | Voltage/Intensity | Soil | Surfactant | Removal (%) | Reference |
|---|---|---|---|---|---|---|
| Mixture of BTEX and PAHs | - | 7.5 V | Clay soil | Cetyltrimenthy lammonium bromide | 97 | (Sri Ranjan et al. 2006) |
| Mixture of PAHs | - | 2 V/cm | Industrial contaminated soil | Tween 80 Igepal CA-720 | | (Reddy et al. 2006) |
| Mixture of PAHs | 10 | 1–2 V/cm | Surficial sediments | Poloxamer 407 Nonidet P40 | 43 48 | (Hahladakis et al. 2013) |
| Mixture of PAHs | - | 1 V/cm | Industrial contaminated soil | Igepal CA-720 | 40 | (Reddy et al. 2010) |
| Mixture of sixteen priority PAHs | 54 | 30–60 V | Marine sediment | Tween 80 | 28 | (Colacicco et al. 2010) |
| Phenanthrene | 600–700 | 10 mA | Kyungnam-Sancheong Kaolin | Alkyl polyglucoside Brij 30 Sodium dodecyl sulfate | 57.8–75.1 45.8–38 44.6–55 | (Park et al. 2007) |
| Phenanthrene | 500–800 | 10 mA | Kaolin | Alkyl polyglucoside Calfax 16L-35 | 63.21–98.03 17.43–24.31 | (Park et al. 2007) |
| Phenanthrene | 100 | 12 V | Natural soil | Triton X-100 Rhamnolipid | 5 20–30 | (Chang et al. 2009) |
| Phenanthrene | 500 | 3 V/cm | Kaolin Sandy soil | Tween 80 Tween 20 Brij 35 | 90 | (Alcántara et al. 2012) |
| Phenanthrene | 500 | 3 V/cm | Kaolin | Triton X-100 Tween 20 | 95 | (Alcántara et al. 2008) |
| Total Petroleum Hydrocarbons | 11,680 | 3 V/cm | Marine sediments | Tween 80 and sodium ethylenediamine tetra acetic | 31 | (Pazos et al. 2013) |

*Electrokinetic treatment with cyclodextrins*

| Pollutant | Conc. (mg/kg) | Voltage/Intensity | Soil | Cyclodextrine | Removal (%) | Reference |
|---|---|---|---|---|---|---|
| Mixture of PAHs | - | 2 V/cm | Industrial contaminated soil | Hydroxypropyl-β-cyclodextrin | - | (Reddy et al. 2006) |
| Phenanthrene | 1.9 | 14 V | Kaolin | Hydroxypropyl-cyclodextrin | 75 | (Ko et al. 2000) |
| Phenanthrene | 500 | 2 V/cm | Kaolin | Hydroxypropyl-cyclodextrin | 50 | (Maturi and Reddy 2006) |

*Electrokinetic treatment with co-solvent*

| Pollutant | Conc. (mg/kg) | Voltage/Intensity | Soil | Co-solvent | Removal (%) | Reference |
|---|---|---|---|---|---|---|
| Benzo[a]pyrene | 300–400 | 3 V/cm | Kaolin | Ethanol | 40 | (Gómez et al. 2009) |
| Mixture of PAHs | - | 2 V/cm | Industrial contaminated soil | n-butylamine | - | (Reddy et al. 2006) |
| Phenanthrene | - | - | Glacial till soil | n-butylamine Tetrahydrofuran Acetone | 43 | (Reddy et al. 2006) |
| Phenanthrene | 500 | 2 V/cm | Kaolin | n-butylamine | 7 | (Maturi and Reddy 2008) |

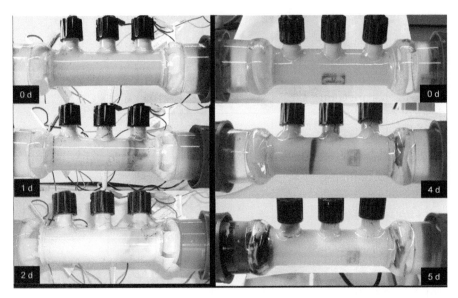

**Fig. 3.** Electrokinetic treatment of kaolin clay polluted with the Lissamine Green B (left) and Reactive Black 5 (right) in an electrokinetic lab-cell operating at electric field of 3 V/cm.

Therefore, Pazos et al. (2008) found that the use of $Na_2HPO_4$ as processing fluid increased the electro-osmotic flow, improved Lissamine Green B desorption from the surface of kaolin, and prevented the acidification of the medium. In Fig. 3 (left) the fast movement of the dye through the clay by the electro-osmotic flow can be appreciated. After two days, 94% of the dye was transported to the cathode chamber where it was accumulated. From there, it can be extracted and easily treated using, for example, electrochemical treatment (Pazos et al. 2008).

Besides the dyes, the electrokinetic treatment has been successfully applied to remove other partially water soluble organic compounds such as pentachlorophenol, hexachlorobenzene, lindane, polycyclic aromatic hydrocarbons (PAH), total petroleum hydrocarbons (TPH), among others (Table 1). In these treatments, the uncharged pollutant is transported by the electro-osmotic flow generated under appropriated conditions. In general, these conditions depend on the soil matrix; therefore an alkaline environment is required in most of the treatments in order to increase the electro-osmotic flow. Therefore, the efficiency of this process is severely limited when the compounds have low solubility. Under these conditions, the electrokinetically enhanced *in situ* flushing technique has the potential to improve soil-solution-contaminant interaction and contaminant removal (Gómez et al. 2009).

### Electrokinetic Treatment with Solubilizing Agents

In the case of organic pollutants with low water solubility, enhancement agents have been used in the electrokinetic remediation process to keep the contaminants in a mobile phase for an appropriated electrokinetic transport. These agents are commonly used to substantially increase organic pollutant desorption and solubilization through micellization and surface tension reduction. During the remediation process, these agents can be added directly into the soil or to the electrode chamber solutions where they are introduced into the soil by electro-osmosis and/or electromigration (Fig. 4). The most common solubilizing agents, utilized in the electrokinetic technique are co-solvents, surfactants, and cyclodextrins.

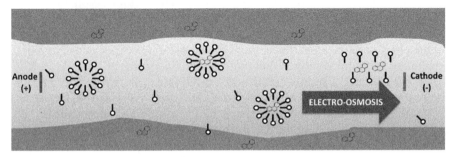

**Fig. 4.** Schematic representation of the PAHs solubilization and migration, by use of surfactants and electric field action, respectively.

### *Co-solvents*

Co-solvents are water-miscible organic compounds as a result of their polar structure. They can promote the removal of organic contaminants in two ways. The first is to increase the apparent solubility of the contaminants in water, which enhances the removal of the mass per unit pore volume. The second way is by reducing the interfacial tension between water and contaminant, which may result in the direct movement of the organic pollutant. Different investigations have shown the high influence of the organic solvent on the adsorption and mobility of organic pollutants (Table 1). Li et al. (2000) performed electrokinetic experiments using different solvents (acetone, tetrahydrofuran, and n-butylamine) to extract phenanthrene from glacial till. They found that after 127 days of treatment the most effective solvent was n-butylamine at a concentration of 20% by removing 43% of the initial phenanthrene. Alcántara et al. (2008) demonstrated that when a solution of 40% ethanol was introduced in the soil by electric field action, the desorption of phenanthrene from kaolin clay was improved and around 95% of initial phenanthrene was removed from the soil and recovered into the cathode chamber.

## Surfactants

Surface active agents (surfactants) are chemical compounds that consist of a strongly hydrophilic group, typically the head of the molecule, and a strongly hydrophobic group which is the tail. The hydrophilic group causes surfactants to exhibit high solubility in water, while the hydrophobic group prefers a hydrophobic phase, such as several organic pollutants. This fact enables surfactants to enhance the solubility of the contaminant through micellar solubilization. In this process, aggregations of surfactant monomers form a micelle; its interior becomes a hydrophobic region suitable for organic pollutants, which improves their solubilization (Alcántara et al. 2008; Gómez et al. 2009; Pazos et al. 2013). Table 1 summarizes a selection of different researches where surfactants were used in order to enhance the decontamination process by the electric field action. Among the different surfactants (anionics, cationics or non-ionics) non-ionics are often chosen in the remediation process for their high solubilization capacity and biodegradability (Choy and Chu 2001). Several investigations proved that non-ionic surfactants (e.g., Igepal CA-720, Tween 80, Tween 20, Triton X-100, Brij 30, Brij 35, Tyloxapol) efficiently improve the removal of PAHs or TPHs by electrokinetic processes (Table 1).

 Gómez et al. (2009) reported that when a solution of surfactant Brij 35 1% was used as processing fluid in the electrokinetic remediation of benzo(a) pyrene, this PAH was successfully transported through kaolin clay towards the cathode chamber. They found that the extent of this recovery depended on the pH profile on the soil. When no pH control was used, around 17% of initial contaminant was detected in the cathode chamber; though, when pH control in the anode chamber was set at 7.0, the recovery of benzo(a)pyrene could be higher than 76%. Similarly, Pazos et al. (2011) used a mixture of Tween 80 and EDTA as processing fluid in the electrokinetic treatment of a soil polluted with diesel fuel. They found that a mixture of both chemical substances increased the removal efficiency reaching around 55% of TPH removal.

## Cyclodextrins

Cyclodextrins are cyclic oligosaccharides of glucopyranose units with a lipophilic cavity in the center. These natural compounds are produced by the action of a group of enzymes known as cyclodextrin glycosyltransferases in starch. Cyclodextrins are capable of forming inclusion complexes with contaminants by taking up a whole contaminant molecule, or some part of it, into the cavity (Gómez et al. 2010). Cyclodextrins appear to be promising agents for improving the solubility of organic compounds because they minimize the environmental impact as a result of their non-toxic nature and biodegradability. The use of these compounds is limited due to their cost and availability; however different researchers have proven their applicability (Table 1). Wang et al. (2013) demonstrated that the electrokinetic remediation combined with glycine-ß-cyclodextrin as processing fluid and pH control may be a good remediation alternative for the treatment of soil polluted with simazine and cadmium. Experimental results reported by Li et al. (2010) showed that migration and removal of hexachlorobenzene in soil was significantly affected by hydroxypropyl-β-cyclodextrin concentrations and cumulative electro-osmotic flow. They found that the electrokinetic remediation combined with hydroxypropyl-β-cyclodextrin flushing and pH buffering was a good alternative for hexachlorobenzene removal from sediments.

## Hybrid Technologies

Different approaches have been developed by coupling electrokinetic technology with other treatments in order to increase the removal efficiency. Most of these technologies use the electric field action to put reagents into the soil to induce the *in situ* degradation of the organic pollutants. In this sense, new technologies called electrokinetic-*Fenton* and electro-bioremediation arise.

### Electrokinetic-Fenton (EK-Fenton)

The use of Fenton's reagent with electrokinetic treatment has been described as an ecological approach for the *in situ* degradation of organic compounds in soils (Kim et al. 2005b). The electrokinetic treatment can facilitate the inclusion of the oxidant ($H_2O_2$) and activates oxidizing radicals, while, at the same time, it may directly cause oxidation or reductive reactions on the soil (Isosaari et al. 2007).

This combined method of oxidation of organic compounds takes place in the soil by means of $H_2O_2$ when iron is naturally present in the soil as catalyst. Hydroxyl radicals generated in the Fenton's reaction are strong oxidants; they oxidize the organic molecule by abstracting hydrogen atoms or by the addition of double bonds and aromatic rings (Iglesias et al. 2013). However, the Fenton reaction is effective only at low pH (about 3), which requires an adjustment of pH during treatment.

The results obtained by several works using this hybrid technology are very encouraging (Table 1). Pazos et al. (2012) demonstrated that using the EK-Fenton process an adequate environment for the Fenton's reaction can be developed inside the soil because the $H_2O_2$ is homogeneously distributed due to the action of the electric field. Their experiences, in the treatment of soil polluted with diesel fuel, reached a removal close to 90%. Furthermore, other chemical substances, as complexing agents, can be introduced in order to increase the removal. Thus, Pazos et al. (2013) treated dredged marine sediments contaminated by petroleum hydrocarbons and metals, by the EK-Fenton treatment. Their research demonstrated that a combination of a Fenton's reagent and sodium ethylenediamine tetra acetic promoted the *in situ* degradation of petroleum hydrocarbons and solubilized the metals. The

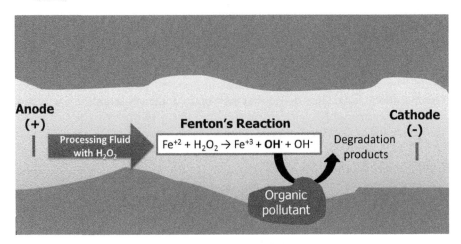

**Fig. 5.** Schematic diagram of EK-Fenton process catalyzed by iron in the soil.

developed EK-Fenton-sodium ethylenediamine tetra acetic process obtained a removal of about 90% for TPH, 57.3% of Zn, 59.8% of Pb, 59.4% of Cu, and 54.5% of Hg.

### Electro-bioremediation

Electro-bioremediation is a promising hybrid technology for the remediation of organic contaminants. Recently, it has been successfully applied in soil with organic contaminants (Table 1). Electrokinetics can be coupled with bioremediation to enhance the removal of organic contaminants, mainly through the potential to mobilize pollutants and bacteria, as well as the ability to stimulate bacterial activity (Table 1). Therefore the electrokinetic treatment can enhance biodegradation by adding nutrients or spreading indigenous bacteria to polluted soil. Jackman et al. (2001) evaluated the electrokinetic movement and biodegradation of 2,4-dichlorophenoxyacetic acid in silt soil. They inoculated the bacterium *Burkholderia* spp. RASC c2 near of the cathode chamber and used the electric field to promote the movement of the pollutant towards the bacteria. At the end of the experiment, 87.1% of radiolabel had been removed from the soil, 5.8% of which was recovered as $^{14}CO_2$. Another way is to transport various additives efficiently to underground polluted area in order to enhance pollutant bioremediation. Accordingly, Pazos et al. (2012) determined that the addition of nutrients via electro-osmotic flow promotes the biostimulation of indigenous bacteria increasing the diesel fuel removal in a loamy soil.

## Conclusions

In this chapter the recent approaches to apply the electrokinetic remediation of soil contaminated with organic pollutants are shown. The main drawback for applying this technology is primarily the low solubility of most organic pollutants. However as it was described in this chapter a wide range of technique improvements can be applied in order to increase the removal efficiency. The application of this technique alone or combined with other treatment processes as, for example, Fenton or bioremediation gives satisfactory results in the removal of organic pollutants.

## Acknowledgements

This research has been financially supported by the Spanish Ministry of Economy and Competitiveness and by ERDF Funds (Project CTM2014-52471-R).

## References

Acar, Y.B. and A.N. Alshawabkeh. 1993. Principles of electrokinetic remediation. Environ. Sci. Technol. 27: 2638–2647.

Alcántara, M.T., J. Gómez, M. Pazos and M.A. Sanromán. 2012. Electrokinetic remediation of lead and phenanthrene polluted soils. Geoderma 173-174: 128–133.

Alcántara, T., M. Pazos, C. Cameselle and M.A. Sanromán. 2008. Electrochemical remediation of phenanthrene from contaminated kaolinite. Environ. Geochem. Health 30: 89–94.

Alshawabkeh, A.N., A.T. Yeung and M.R. Bricka. 1999. Practical aspects of *in situ* electrokinetic extraction. J. Environ. Eng. 125: 27–35.

Cang, L., G.P. Fan, D.M. Zhou and Q.Y. Wang. 2013. Enhanced-electrokinetic remediation of copper-pyrene co-contaminated soil with different oxidants and pH control. Chemosphere 90: 2326–2331.

Chang, J.H., Z. Qiang, C.P. Huang and A.V. Ellis. 2009. Phenanthrene removal in unsaturated soils treated by electrokinetics with different surfactants-Triton X-100 and rhamnolipid. Colloids Surf. A Physicochem. Eng. Asp. 348: 157–163.

Choy, W.K. and W. Chu. 2001. The modelling of trichloroethene photo degradation in Brij 35 surfactant by two-stage reaction. Chemosphere 44: 211–215.

Colacicco, A., G. De Gioannis, A. Muntoni, E. Pettinao, A. Polettini and R. Pomi. 2010. Enhanced electrokinetic treatment of marine sediments contaminated by heavy metals and PAHs. Chemosphere 81: 46–56.

Dong, Z.Y., W.H. Huang, D.F. Xing and H.F. Zhang. 2013. Remediation of soil co-contaminated with petroleum and heavy metals by the integration of electrokinetics and biostimulation. J. Hazard. Mater. 260: 399–408.

Fernandez, A., P. Hlavackova, V. Pomès and M. Sardin. 2009. Physico-chemical limitations during the electrokinetic treatment of a polluted soil. Chem. Eng. J. 145: 355–361.

Gómez, J., M.T. Alcántara, M. Pazos and M.A. Sanromán. 2009. A two-stage process using electrokinetic remediation and electrochemical degradation for treating benzo[a]pyrene spiked kaolin. Chemosphere 74: 1516–1521.

Gómez, J., M.T. Alcántara, M. Pazos and M.A. Sanromán. 2010. Soil washing using cyclodextrins and their recovery by application of electrochemical technology. Chem. Eng. J. 159: 53–57.

Gonzini, O., A. Plaza, L. Di Palma and M.C. Lobo. 2010. Electrokinetic remediation of gasoil contaminated soil enhanced by rhamnolipid. J. Appl. Electrochem. 40: 1239–1248.

Hahladakis, J.N., A. Lekkas, A. Smponias and E. Gidarakos. 2014. Sequential application of chelating agents and innovative surfactants for the enhanced electroremediation of real sediments from toxic metals and PAHs. Chemosphere 105: 44–52.

Huang, D., Q. Xu, J. Cheng, X. Lu and H. Zhang. 2012. Electrokinetic remediation and its combined technologies for removal of organic pollutants from contaminated soils. Int. J. Electrochem. Sci. 7: 4528–4544.

Iglesias, O., M.A. Fernández de Dios, M. Pazos and M.A. Sanromán. 2013. Using iron-loaded sepiolite obtained by adsorption as a catalyst in the electro-Fenton oxidation of Reactive Black 5. Environ. Sci. Pollut. Res. 20: 5983–5993.

Isosaari, P., R. Piskonen, P. Ojala, S. Voipio, K. Eilola, E. Lehmus and M. Itävaara. 2007. Integration of electrokinetics and chemical oxidation for the remediation of creosote-contaminated clay. J. Hazard. Mater. 144: 538–548.

Jackman, S.A., G. Maini, A.K. Sharman, G. Sunderland and C.J. Knowles. 2001. Electrokinetic movement and biodegradation of 2,4-dichlorophenoxyacetic acid in silt soil. Biotechnol. Bioeng. 74: 40–48.

Kim, B.K., K. Baek, S.H. Ko and J.W. Yang. 2011. Research and field experiences on electrokinetic remediation in South Korea. Sep. Purif. Technol. 79: 116–123.

Kim, J.H., S.S. Kim and J.W. Yang. 2007. Role of stabilizers for treatment of clayey soil contaminated with phenanthrene through electrokinetic-Fenton process-Some experimental evidences. Electrochim. Acta 53: 1663–1670.

Kim, S.J., J.Y. Park, Y.J. Lee, J.Y. Lee and J.W. Yang. 2005a. Application of a new electrolyte circulation method for the *ex situ* electrokinetic bioremediation of a laboratory-prepared pentadecane contaminated kaolinite. J. Hazard. Mater. 118: 171–176.

Kim, S.S., J.H. Kim and S.J. Han. 2005b. Application of the electrokinetic-Fenton process for the remediation of kaolinite contaminated with phenanthrene. J. Hazard. Mater. 118: 121–131.

Ko, S.O., M.A. Schlautman and E.R. Carraway. 2000. Cyclodextrin-enhanced electrokinetic removal of phenanthrene from a model clay soil. Environ. Sci. Technol. 34: 1535–1541.

Lee, H.H. and J.W. Yang. 2000. A new method to control electrolytes pH by circulation system in electrokinetic soil remediation. J. Hazard. Mater. 77: 227–240.

Li, A., K.A. Cheung and K.R. Reddy. 2000. Cosolvent-enhanced electrokinetic remediation of soils contaminated with phenanthrene. J. Environ. Eng. 126: 527–533.

Li, F., S. Guo and N. Hartog. 2012. Electrokinetics-enhanced biodegradation of heavy polycyclic aromatic hydrocarbons in soil around iron and steel industries. Electrochim. Acta. 85: 228–234.

Li, T., S. Guo, B. Wu, F. Li and Z. Niu. 2010. Effect of electric intensity on the microbial degradation of petroleum pollutants in soil. J. Environ. Sci. 22: 1381–1386.

Maturi, K. and K.R. Reddy. 2006. Simultaneous removal of organic compounds and heavy metals from soils by electrokinetic remediation with a modified cyclodextrin. Chemosphere 63: 1022–1031.

Maturi, K. and K.R. Reddy. 2008. Cosolvent-enhanced desorption and transport of heavy metals and organic contaminants in soils during electrokinetic remediation. Water Air Soil Pollut. 189: 199–211.

Murillo-Rivera, B., I. Labastida, J. Barrón, M.T. Oropeza-Guzman, I. González and M.M.M. Teutli-Leon. 2009. Influence of anolyte and catholyte composition on TPHs removal from low permeability soil by electrokinetic reclamation. Electrochim. Acta 54. 2119–2124.

Niqui-Arroyo, J.-L., M. Bueno-Montes, R. Posada-Baquero and J.J. Ortega-Calvo. 2006. Electrokinetic enhancement of phenanthrene biodegradation in creosote-polluted clay soil. Environ. Pollut. 142: 326–332.

Niroumand, H., R. Nazir and K.A. Kassim. 2012. The performance of electrochemical remediation technologies in soil mechanics. Int. J. Electrochem. Sci. 7: 5708–5715.

Park, J.Y. and J.H. Kim. 2011. Switching effects of electrode polarity and introduction direction of reagents in electrokinetic-Fenton process with anionic surfactant for remediating iron-rich soil contaminated with phenanthrene. Electrochim. Acta 56: 8094–8100.

Park, J.Y., S.J. Kim, Y.J. Lee, K. Baek and J.W. Yang. 2005. EK-Fenton process for removal of phenanthrene in a two-dimensional soil system. Eng. Geol. 77: 217–224.

Park, J.Y., H.H. Lee, S.J. Kim, Y.J. Lee and J.W. Yang. 2007. Surfactant-enhanced electrokinetic removal of phenanthrene from kaolinite. J. Hazard. Mater. 140: 230–236.

Pazos, M., M.A. Sanromán and C. Cameselle. 2006. Improvement in electrokinetic remediation of heavy metal spiked kaolin with the polarity exchange technique. Chemosphere 62: 817–822.

Pazos, M., M.T. Ricart, M.A. Sanromán and C. Cameselle. 2007. Enhanced electrokinetic remediation of polluted kaolinite with an azo dye. Electrochim. Acta 52: 3393–3398.

Pazos, M., C. Cameselle and M.A. Sanromán. 2008. Remediation of dye-polluted kaolinite by combination of electrokinetic remediation and electrochemical treatment. Environ. Eng. Sci. 25: 419–428.

Pazos, M., M.T. Alcántara, C. Cameselle and M.A. Sanromán. 2009. Evaluation of electrokinetic technique for industrial waste decontamination. Sep. Sci. Technol. 44: 2304–2321.

Pazos, M., M.T. Alcántara, E. Rosales and M.A. Sanromán. 2011. Hybrid Technologies for the Remediation of Diesel Fuel Polluted Soil. Chem. Eng. Technol. 34: 2077–2082.

Pazos, M., A. Plaza, M. Martín and M.C. Lobo. 2012. The impact of electrokinetic treatment on a loamy-sand soil properties. Chem. Eng. J. 183: 231–237.

Pazos, M., O. Iglesias, J. Gómez, E. Rosales and M.A. Sanromán. 2013. Remediation of contaminated marine sediment using electrokinetic-Fenton technology. J. Ind. Eng. Chem. 19: 932–937.

Pham, T.D., R.A. Shrestha, J. Virkutyte and M. Sillanpää. 2009. Combined ultrasonication and electrokinetic remediation for persistent organic removal from contaminated kaolin. Electrochim. Acta. 54: 1403–1407.

Probstein, R.F. and R.E. Hicks. 1993. Removal of contaminants from soils by electric fields. Science 260: 498–503.

Reddy, K.R., P.R. Ala, S. Sharma and S.N. Kumar. 2006. Enhanced electrokinetic remediation of contaminated manufactured gas plant soil. Eng. Geol. 85: 132–146.

Reddy, K.R., K. Maturi and C. Cameselle. 2009. Sequential electrokinetic remediation of mixed contaminants in low-permeability soils. J. Environ. Eng. 135: 989–998.

Reddy, K.R., C. Cameselle and P. Ala. 2010. Integrated electrokinetic-soil flushing to remove mixed organic and metal contaminants. J. Appl. Electrochem. 40: 1269–1279.

Reddy, K.R., K. Darko-Kagya and A.Z. Al-Hamdan. 2011a. Electrokinetic remediation of chlorinated aromatic and nitroaromatic organic contaminants in clay soil. Environ. Eng. Sci. 28: 405–413.

Reddy, K.R., K. Darko-Kagya and A.Z. Al-Hamdan. 2011b. Electrokinetic remediation of pentachlorophenol contaminated clay soil. Water Air Soil Pollut. 221: 35–44.

Roriz, M.S., J.F. Osma, J.A. Teixeira and S. Rodríguez Couto. 2009. Application of response surface methodological approach to optimise Reactive Black 5 decolouration by crude laccase from *Trametes pubescens*. J. Hazard. Mater. 169: 691–696.

Saichek, R.E. and K.R. Reddy. 2003. Effect of pH control at the anode for the electrokinetic removal of phenanthrene fromkaolin soil. Chemosphere 51: 273–287.

Sanromán, M.A., M. Pazos, M.T. Ricart and C. Cameselle. 2005. Decolourisation of textile indigo dye by DC electric current. Eng. Geol. 77: 253–261.

Sri Ranjan, R., Y. Qian and M. Krishnapillai. 2006. Effects of electrokinetics and cationic surfactant cetyltrimenthylammonium bromide [CTAB] on the hydrocarbon removal and retention from contaminated soils. Environ. Technol. 27 : 767–776.

Tsai, T.T., J. Sah and C.M. Kao. 2010. Application of iron electrode corrosion enhanced electrokinetic-Fenton oxidation to remediate diesel contaminated soils: A laboratory feasibility study. J. Hydrol. 380: 4–13.

Wan, C., M. Du, D.J. Lee, X. Yang, W. Ma and L. Zheng. 2011. Electrokinetic remediation of ß-cyclodextrin dissolved petroleum hydrocarbon-contaminated soil using multiple electrodes. J. Taiwan Inst. Chem. Eng. 42: 972–975.

Wang, G., J. Luo and N. Deng. 2013. Enhancement of electrokinetic removal of simazine and cadmium from co-contaminated soils by glycine-ß-cyclodextrin. Fresenius Environ. Bull. 22: 1904–1912.

Wieczorek, S., H. Weigand, M. Schmid and C. Marb. 2005. Electrokinetic remediation of an electroplating site: Design and scale-up for an *in-situ* application in the unsaturated zone. Eng. Geol. 77: 203–215.

Xu, W., C. Wang, H. Liu, Z. Zhang and H. Sun. 2010. A laboratory feasibility study on a new electrokinetic nutrient injection pattern and bioremediation of phenanthrene in a clayey soil. J. Hazard. Mater. 184: 798–804.

Yang, G.C.C. and C.Y. Liu. 2001. Remediation of TCE contaminated soils by *in situ* EK-Fenton process. J. Hazard. Mater. 85: 317–331.

Yang, J.W., Y.J. Lee, J.Y. Park, S.J. Kim and J.Y. Lee. 2005. Application of APG and Calfax 16L-35 on surfactant-enhanced electrokinetic removal of phenanthrene from kaolinite. Eng. Geol. 77: 243–251.

Yu, J.W. and I. Neretnieks. 1996. Modelling of transport and reaction processes in a porous medium in an electrical field. Chem. Eng. Sci. 51: 4355–4368.

# CHAPTER 2

# Bioremediation of Chlorinated Ethenes

*Anthony S. Danko*[1,*] *and James M. Cashwell*[2]

## ABSTRACT

Chlorinated ethenes have been widely used in an array of industries, such as those involved in degreasing and dry cleaning. Improper use, handling, and disposal of these compounds have ultimately led to widespread groundwater contamination in Europe, the US, and elsewhere. Since exposure to these contaminants may cause unacceptable health risks, significant work has been done to gain insights into the remediation of these compounds from a scientific and applications point of view. This chapter focuses on the different ways in which chlorinated ethenes are biodegraded in the environment. In addition, practical implementation of different techniques is also discussed, including the addition of microorganisms and different biostimulants, dosing requirements, and delivery options.

## Introduction

Chlorinated ethenes, such as perchloroethene (PCE; $C_2Cl_4$), and trichloroethene (TCE; $C_2HCl_3$), were typically used in a variety of industries as degreasing agents and dry cleaning solvents. This was due to their physical and chemical properties, as well as their low cost. Their extensive use combined

[1] Centre for Natural Resources and the Environment (CERENA), Department of Mining Engineering, University of Porto - Faculty of Engineering (FEUP), Rua Dr. Roberto Frias s/n, 4200-465, Porto, Portugal.
[2] Olin Corporation, 3855 N. Ocoee Street, Suite 200, Cleveland, TN, United States of America 37312.
* Corresponding author: asdanko@fe.up.pt

with accidental releases and improper treatment and disposal have led to widespread contamination in the environment.

Once spilled, these compounds may migrate through the vadose zone and penetrate the groundwater. At high enough concentrations, PCE and TCE may exist as dense, non-aqueous phase liquids (DNAPLs), providing a source of contamination for years to come. This is due to their limited aqueous solubility and the fact that they are denser than water (EPA 2004). In addition, due to their volatility (EPA 2004), chloroethene vapors may percolate into the vadose zone and contaminate indoor air in homes and/or buildings.

Exposure to these chemicals may cause unacceptable risks to human health due to their toxicity (IARC 1995). Indeed, PCE and TCE are suspected carcinogens while a breakdown product, vinyl chloride ($C_2H_3Cl$; VC), is a known carcinogen. Due to this, low exposure limits have been mandated by a variety of regulatory bodies for exposure to air or water where the presence of these compounds is detected.

Various physical and chemical approaches, including soil vapor extraction, zero-valent iron, permanganate, surfactant addition, etc., have been used to clean up groundwater contaminated with PCE and TCE. Initially, pump and treat was the preferred method. However, because most of the subsurface mass of PCE and TCE is adsorbed to soil and the rate of desorption is slow, long pumping times are usually required for mass removal (Cowell et al. 2000; Shimotori and Arnold 2002). The technologies described above are often times inefficient or not cost effective at a variety of sites. Therefore, the development of other technologies, especially *in situ* bioremediation approaches, may offer an attractive alternative. *In situ* bioremediation offers several advantages over physical and chemical methods of treatment. Generally, *in situ* bioremediation has lower operational and maintenance costs and offers the potential to transform the contaminants to non-hazardous products (Freedman and Gossett 1989; Becker and Freedman 1994).

## Biodegradation of Chlorinated Ethenes

Due to the widespread presence of PCE and TCE in the environment, combined with its toxic nature, there has been intense interest in the development of *in situ* bioremediation technologies and the identification and characterization of microorganisms that transform these compounds. There are many microorganisms that have been found that carry out these reactions and can be classified into four main groups: anaerobic reductive dechlorination, anaerobic oxidation, aerobic metabolism, and aerobic cometabolism. Each group will be described in the following sections.

### Anaerobic Reductive Dechlorination of Chlorinated Ethenes

Many microorganisms are capable of reducing PCE and TCE into lesser chlorinated products under anaerobic conditions. The complete dechlorination of PCE or TCE leads to ethene, although ethane can also be produced. In addition, the predominant dichloroethene isomer produced from reductive dechlorination is *cis*-DCE; however, *trans*-DCE has also been found (Griffin et al. 2004).

This biotransformation process, also known as reductive dechlorination, uses hydrogen ($H_2$) as the electron donor while releasing $Cl^-$. Each dechlorination step produces one $H^+$ and one $Cl^-$ while consuming $2H^+$ and $2e^-$. As a result of this process, the organic contaminant becomes reduced. The reductive dechlorination pathway for PCE is shown in Fig. 1.

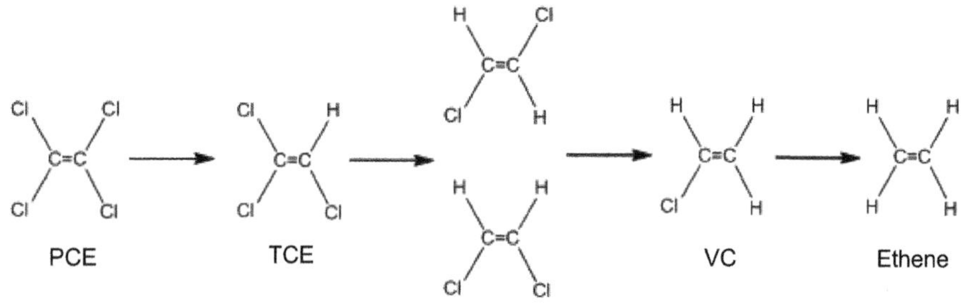

**Fig. 1.** Reductive Dechlorination of PCE.

Examples of microorganisms that can reduce PCE or TCE include methanogens, *Acetobacterium woodii*, and *Clostridium bifermentans* (Vogel and McCarty 1985; Fathepure et al. 1987; Terzenbach and Blaut 1994; Chang et al. 2000); however, they can generally only partially reduce PCE and TCE to lesser dechlorination products (such as *cis*-DCE) via a process called cometabolism. Cometabolism is a non-growth linked biodegradation reaction in which the microorganisms fortuitously transform the contaminant of interest (aerobic cometabolism via oxidation is discussed in section "Aerobic Cometabolism of Chlorinated Ethenes"). Some dechlorinating microorganisms, such as *Dehalobacter restrictus* and *Sulfurospirillum multivorans*, also partially biodegrade PCE and TCE to *cis*-DCE but utilize a growth linked process, which is also known as halorespiration. This means that they are able to gain energy from the process.

The partial transformation of PCE and TCE can be problematic if the reduction stops at *cis*-DCE and VC. VC is often the regulatory driver at sites because it is a known carcinogen and has the lowest maximum contaminant level of any volatile organic compound (Pontius 1996). Therefore, microorganisms capable of the complete reduction of PCE or TCE to ethene are of great interest.

Recently, microorganisms of the genus *Dehalococcoides* (*Dhc*) were found to completely reduce PCE and TCE to ethane (Löffler et al. 2013). The first microorganism discovered to do this was *Dehalococcoides mccartyi* 195 (formerly *Dehalococcoid esethenogenes* 195) (Maymó-Gatell et al. 1997). Characterization of the strain demonstrated that strain 195 metabolically reduced PCE to VC and cometabolically reduced VC to ethene. Subsequently, many other *Dhc* isolates were found to carry out this reaction, including those that can metabolically reduce VC to ethene. *Dhc* isolates that are capable of respiring select chlorinated ethenes include strains BAV1, FL2, GT, VS, and CBDB1. Care should be used because each *Dhc* strain has its own electron acceptor range and which electron acceptors can be utilized metabolically and cometabolically (Maymó-Gatell et al. 1997; Cupples et al. 2003; He et al. 2003b; Sung et al. 2006; Marco-Urrea et al. 2011).

*Dhc* is normally cultured in a consortium with other microorganisms, such as methanogens and acetogens. This is because it is notoriously difficult to isolate and grow since they have doubling times of 1 to 2 days (Maymó-Gatell et al. 1997; He et al. 2003a; He et al. 2003b; Sung et al. 2006; Marco-Urrea et al. 2011) and complex nutrient requirements. These consortia allow for a more robust dechlorination since the microbial community not only provides trace nutrients and $H_2$ (via the fermentation of lactate or other carbon sources) for *Dhc*, but it also balances Eh, redox conditions, and buffers against other potentially inhibitory compounds. Methods of culturing *Dhc* in pure culture and in mixed cultures can be found in these references (Vainberg and Steffan 2009; Löffler et al. 2013).

Sites where incomplete dechlorination occurs could be the result of a variety of reasons, including the lack of or presence of *Dhc* in low numbers at the site, improper electron accepting conditions, or the presence of inhibitory compounds. Sites which lack the presence of *Dhc* can be cleaned up using bioaugmentation, which is the exogenous addition of cultures that contain *Dhc*. Biostimulation with lactate, or the addition of exogenous electron donor, can be used to ensure proper redox conditions are present. Engineering aspects about biostimulation and bioaugmentation are found later in this chapter (Sections "*In Situ* Application Techniques—Biostimulation" and "Solid Biostimulants").

Because *Dhc* is necessary for the complete reduction to ethene, molecular methods have been developed to detect this microorganism. However, similarities between the 16S ribosomal nucleic acid (rRNA) sequences among *Dhc* strains and the metabolic differences between them limit their efficacy in monitoring. Most remediation practitioners combine the application of

16S-rRNA detection of *Dhc* by also monitoring functional dehalogenase genes. These genes include pceA (PCE → TCE), tceA (TCE → VC), vcrA (DCEs and VC → ethene), and bvcA (VC → ethene).

### Anaerobic Oxidation of Chlorinated Ethenes

Whereas in reductive dechlorination, the chlorinated compound is the electron acceptor and becomes reduced, in anaerobic oxidation the chlorinated compound is oxidized, rather than reduced. Anaerobic oxidation has been observed for *cis*-DCE and VC under a variety of redox conditions. These include iron (III) reducing, manganese (IV) reducing, sulfate reducing, and methanogenic conditions (Bradley and Chapelle 1996, 1997; Bradley et al. 1998). In addition, microcosms incubated with TCE from a site that contained fractured sandstone suggested that the *cis*-DCE that was anaerobically produced from TCE was oxidized to organic acids and $CO_2$ by the presence of reactive iron minerals (pyrite, etc.) in an abiotic reaction (Darlington et al. 2008). Anaerobic oxidation is very difficult to measure due to the production of these acids and/or carbon dioxide. Indeed, almost all documented work within this subject has utilized $^{14}C$ in order to confirm the products of (bio)degradation. In addition, recent evidence has shown that aerobic biodegradation of VC can occur even at dissolved oxygen concentrations of < 0.1 mg/L (Gossett 2010). This may mean that at some sites, the loss of VC suspected via anaerobic oxidation may have actually been occurring by aerobic biodegradation by oxygen diffusion from nearby aerobic regions and/or recharge (Gossett 2010). In any event, there are many unknowns associated with anaerobic oxidation, including the microorganisms involved (and genes), mechanisms, and geochemical interactions. This has made this process difficult to use from a remediation perspective.

### Aerobic Metabolism of Chlorinated Ethenes

Many microorganisms are capable of catabolizing VC and ethene. However, there has only been one microorganism capable of growth on *cis*-DCE as the sole carbon and energy source (Coleman et al. 2002a). In addition, a mixed culture was recently shown to metabolize TCE (Schmidt et al. 2014). Growth on PCE has not been observed. Microorganisms involved in the aerobic metabolism of chlorinated ethenes include several strains of *Mycobacterium* and *Pseudomonas*, and one strain each of *Ochrobactrum*, *Nocardioides*, and *Ralstonia* (Verce et al. 2000; Coleman et al. 2002b; Danko et al. 2004; Elango et al. 2006). Although, the disappearance of VC is observed at many sites, it is by no means ubiquitous. Indeed, several studies have shown that VC biodegradation activity ranged between 30% to 50% (Coleman et al. 2002b; Madl 2002). Therefore, care should be taken in the assumption that VC will be biodegraded at sites.

VC biodegradation is initiated by a Monooxygenase, producing the toxic intermediate VC-epoxide. Recently, it was discovered that VC catabolic bacteria use epoxyalkane-coenzyme M transformase (EaCoMT) in order to transform the epoxide into a non-ringed structure (Coleman and Spain 2003a,b; Mattes et al. 2005; Danko et al. 2006). The discovery of the fact that coenzyme M (CoM) was the cofactor utilized by the bacteria was interesting, since CoM is used by methanogenic bacteria in the production of methane. Beyond the use of CoM, the pathway for VC catabolism is less known; however, it is generally assumed that the pathway leads to central metabolism. Genes associated with the catabolism of VC are associated with plasmids (Coleman and Spain 2003a; Danko et al. 2004; Mattes et al. 2005). Molecular tools are available that can be utilized to test for the presence of the Monooxygenase and EaCoMT genes.

However, care must be used since the EaCoMT genes are found in both ethene and VC catabolic microorganisms and not all etheneothrophs can grow on VC. Therefore, the detection of biomarkers, including polypeptides using peptide mass fingerprinting (Chuang and Mattes 2007) is being developed since it may be useful in delineating between the different catabolic capabilities of the different microbial populations.

Catabolism of *cis*-DCE as the sole carbon and energy source has so far only been found in one microorganism—*Polaromonas* JS666 (Coleman et al. 2002a). A lot of work has been carried out to identify the pathway strain JS666 utilizes. Evidence suggests that this strain uses a monoxygenase to initiate the reaction producing *cis*-DCE epoxide. Further steps involve multiple reactions and different pathways (one possibly via glutathione-S-transferase) but eventually lead to glyoxalate (Jennings et al. 2009; Nishino et al. 2013). Like VC, genes associated with the biodegradation of *cis*-DCE are associated with plasmids (Mattes et al. 2008).

Since this is the only microorganism capable of *cis*-DCE catabolism, there is significant interest in further characterization of it for bioaugmentation purposes. JS666 was successfully tested in microcosms from sites contaminated with *cis*-DCE (Giddings et al. 2010). In addition, the isocitratelyase gene has been examined for its usefulness in monitoring (Giddings et al. 2010). Work is ongoing concerning the application of the strain and molecular tools for monitoring JS666 at a full scale site.

Recently, a mixed microbial culture has been shown to aerobically metabolize TCE without the addition of a primary substrate (Schmidt et al. 2014). This culture and information are of significant interest due to the amount of contaminated sites that contain TCE. There are potentially several advantages using this approach since biodegradation can occur without anaerobic conditions (electron donor addition) and the carcinogen VC is not formed during this process. However, much like anaerobic oxidation, further research is need in order to further characterize the process, including the identification of the microorganisms involved (and genes), pathways, bioaugmentation potential, impacts of groundwater characteristics (pH, etc.) and concentrations of TCE and other potential co-contaminants on biodegradation.

### Aerobic Cometabolism of Chlorinated Ethenes

As stated in Section "Anaerobic Oxidation of Chlorinated Ethenes", cometabolism is a non-growth linked biodegradation reaction in which the microorganisms fortuitously transform the contaminant of interest. Therefore, the primary substrate must be supplied at some point to stimulate growth of the microorganisms. Primary substrates that drive cometabolism include a variety of alkenes and alkanes, such as methane, ethene, ethane, propene, propane, butane, etc. (Alvarez-Cohen and Speitel 2001). Other compounds, such as ammonia and aromatic hydrocarbons (benzene, toluene, etc.) have also shown to be effective for aerobic cometabolism. Alkanes and alkenes are advantageous due to their non-toxic nature and the fact that they are gases, which allows for better control on delivery while the later primary substrates (ammonia, aromatics) may be considered toxic and limits their use in the field for active remediation but may be useful in natural attenuation.

The enzymes involved in aerobic cometabolism are very robust in the transformation of chlorinated compounds and they are used by the microorganism to initiate transformation of their growth substrate. In this case, microorganisms that utilize an 'oxygenase' enzyme are able to cometabolize a variety of chlorinated ethenes, including VC, *cis*-DCE, and TCE; PCE is not transformed (Alvarez-Cohen and Speitel 2001). However, there have been reports of fungi and enzymes that have been genetically engineered that have this capability but their applications may be limited.

Aerobic cometabolism has successfully been implemented in the field (Alvarez-Cohen and Speitel 2001; Frascari et al. 2015). This includes a variety of different technologies, including air sparging alone; air sparging plus growth substrate; groundwater extraction, amendment, and re-injection, etc. (Frascari et al. 2015). Care must be taken to ensure the effective addition of primary substrate and oxygen since the constant addition of primary substrate may lead to clogging near the injection well or cause a very small biotreatment zone due to non-uniform distribution of biomass. Therefore, pulsed feeding of primary substrate and oxygen may be warranted. In addition, bioaugmentation may also be used for *in situ* cometabolism for a variety of reasons. These include decreased lag time compared to biostimulation, reduced risks of aquifer clogging due to biomass growth, and reduced risks of CAH volatilization due to sparging, etc. (Frascari et al. 2015).

Like other approaches, molecular methods are available for aerobic cometabolism to measure specific microbial populations and oxygenase genes to assist in the monitoring of active remediation.

### *In Situ Application Techniques—Biostimulation*

The metabolic variations described above (i.e., reductive dechlorination, aerobic metabolism, etc.) have one common over-arching requirement: subsurface geochemical conditions must be appropriate to facilitate and maintain the reactions that result in the desired degradation and/or transformation. Appropriate geochemical conditions must be created or enhanced in many cases for *in situ* bioremediation to be successful. For instance, as mentioned in section "Anaerobic Reductive Dechlorination of Chlorinated Ethenes", reductive dechlorination of PCE requires an input of 2 electrons per mole of PCE under strictly anaerobic conditions to degrade PCE to TCE. The input of electrons must come from an appropriate electron donor such as hydrogen ($H_2$). In turn, a source of $H_2$, lactate for instance, must be present to sustain long-term system effectiveness. Biostimulants such as lactate must be delivered appropriately to provide the required electron input in cases where the naturally-occurring dissolved organic carbon concentrations in the targeted remediation area are too low to provide the stoichiometrically required mass of $H_2$.

Field delivery of biostimulants to impacted subsurface areas can be accomplished in a variety of ways. This section provides a summary of common biostimulants and various common *in situ* delivery techniques used to deliver these chemicals to the subsurface environment. The techniques described herein are based on the experience of the author, are *typically* used, and do not constitute an exhaustive or *complete* summary of techniques available. Many of the delivery techniques in the remediation industry are patented in the United States and various other countries. The discussion provided herein is intended to give an overview of general techniques with an attempt to avoid unintentional claims of intellectual ownership or to inadvertently take credit for the invention of the techniques themselves. In addition, it should be noted that it is not the author's intent to promote any products preferentially, but rather to note typical products that are commonly used in the industry.

Chemicals that serve as biostimulants, or amendments as described above, exist in solid, liquid, and gas phases. The associated delivery techniques will largely depend on the phase of the chemicals being delivered.

### Solid biostimulants

Many different types of solid and/or semi-solid amendments have been developed over recent years to enhance aerobic and/or anaerobic degradation of various contaminants. For example, ORC® has been developed by Regenesis (San Clemente, CA, USA) to serve as a controlled-release of oxygen. Regenesis' website (www.Regenesis.com) defines ORC® or Oxygen Release Compound as "...a proprietary formulation of phosphate-intercalated magnesium peroxide that, when hydrated, produces a controlled release of oxygen for periods of up to 12 months on a single application." Other metal peroxides are also on the market. FMC (Philadelphia, PA, USA) has developed Permeox® Plus which is a form of calcium peroxide that is intended also to provide a slow release of oxygen into the subsurface to enhance aerobic degradation. FMC (http://environmental.fmc.com) describes Permeox® Plus as "...a specially formulated time-release grade of calcium peroxide designed to assist in the aerobic bioremediation of hydrocarbons in soil and groundwater."

Various types of mulch, compost, etc. have been used to enhance anaerobic degradation of chlorinated solvents. Amendments such as processed corn cobs, eucalyptus mulch, and commercial compost products were successful in degrading TCE (Katsenovich et al. 2007). The ESTCP (Environmental Security Technology Certification Program—United States Department of Defense) published a Cost and Performance Report (Department of Defense 2008) associated with the use of mulch biowalls for treatment of RDX and/or HMX. The study showed that mulch walls can offer an effective means of treatment of shallow saturated zones with cost advantages over other similar technologies that rely on emplacement of amendments rather than pressurized injection.

Zero-valent iron (ZVI) in various forms has been used extensively over the last several years to enhance anaerobic conditions for *in situ* treatment of chlorinated ethenes. ZVI combined with emulsified vegetable oil creates strong anaerobic conditions while providing a source of electrons that will serve to maintain appropriate oxidation/reduction conditions.

These types of biostimulants exist in the solid phase when procured. In the case of metal hydroxides (ORC® or Permeox®-Plus), they are powder-like and are typically used to create slurries that can be injected under pressure or emplaced directly into areas that have been excavated or other subsurface features such as trenches, pits, etc. It is important that slurries of these compounds be stirred or agitated throughout delivery to maintain the slurry. These biostimulants can also be contained in mesh sleeves, or socks, that can be placed into typical monitoring wells for dissolution and distribution by advective, dispersive, and/or diffusive mass transport.

The various types of mulch can be procured in bulk from local nurseries or home improvement businesses. However, the volume of mulch needed can often exceed typical commercial supplies such that finding a local source of mulch in the desired quantities can be one of the more significant challenges to this technology.

ZVI may be procured from a manufacturer specializing in production of this material (e.g., Hepure Technologies, Inc.—Wilmington, Delaware, USA; Carus Corporation—Peru, IL; etc.). More typically, ZVI application is implemented by a vendor that specializes in remediation using the various forms of ZVI (e.g., SiREM, Guelph, Ontario, Canada; Geosierra Environmental, Inc.—Norcross, Georgia, USA). As will be discussed below, the application of ZVI has become a technical specialty that relies on significant experience and testing which is shared by a number of remediation contractors.

## Liquid biostimulants

There are many different liquid biostimulants that can be used to enhance or create either aerobic or anaerobic degradation/transformation. The list of biostimulants used to facilitate anaerobic reactions is much longer than those for the aerobic case due to the variety of metabolic requirements associated with the wide range of anaerobic microorganisms.

From the aerobic standpoint, dilute hydrogen peroxide has been used quite effectively by the author to enhance/create aerobic conditions. It is advised to take great care when working with hydrogen peroxide at high concentrations (i.e., 20–50%) given the oxidative strength of the compound. It is often more efficient to procure higher concentrations of this compound and dilute it to the required final concentration given the volumes typically required.

From an anaerobic standpoint, both proprietary and commercially-available compounds have been used effectively to create and maintain anaerobic conditions *in situ*. Most of these compounds are derivatives of simple organic acids such as lactate, acetate, citrate, or in many cases the organic acids themselves. Proprietary versions include compounds such as HRC® (Hydrogen Release Compound®—Regenesis). HRC® is comprised of three molecules of lactate esterified to a glycerol backbone (glycerol polylactate). Some vendors have patented the use of common compounds as biostimulants for *in situ* biodegradation (e.g., Arcadis' patented use of molasses and a source of simple sugars). Other common compounds such as vegetable oil have become a readily-available source of carbohydrates that can be used in various forms as a biostimulant.

The key to success for using any of these compounds, whether it is for enhancement of aerobic or anaerobic conditions, is delivery of the specific biostimulants to the area of contamination in amounts that will satisfy the total electron demand. The following sections describe the dosing requirements and typical delivery methods for use of the compounds described herein.

## Gaseous biostimulants

One of the most common gaseous biostimulants is simply air for enhancement of aerobic conditions. The contaminants that are aerobically degraded are often volatile so the rate of air injection is an important design consideration. In the author's experience, flow rates beyond 0.5 standard cubic foot per minute (0.014 cubic meters/minute) are likely excessive and can serve to volatilize the contaminant and strip them out of the subsurface rather than result in degradation.

Hydrogen and oxygen have also been used as injectates. Obviously, use of these flammable substances must be considered with caution. These compounds are typically introduced into the subsurface environment via diffusive techniques (Waterloo emitter™, KerfootMicroporous diffuser, etc.)

which create significant concentration gradients that result in 3-dimensional diffusion according to Fick's law (described below).

$$J = -D\nabla\phi$$

Where, J = 3-dimensional diffusive flux (mol/m$^{2*}$sec); D = diffusion coefficient (m$^2$/sec); and φ = concentration (mol/m$^3$).

## *Dosing requirements*

A thorough understanding of the geochemistry that characterizes any given contaminated zone is essential to determining the appropriate biostimulant dosage. This includes a reasonable understanding of the total mass of contaminant in the target area. It must be understood that the concentrations of contaminants in a groundwater sample do not typically represent the total amount of contaminant present in the subsurface. Most contaminants will adsorb to the soil and/or rock matrix to varying degrees according to corresponding sorption coefficients ($K_d$). This sorbed portion of the total mass will eventually desorb in such a way as to maintain equilibrium between the soil and aqueous phases. There is also the significant potential that a portion of the contaminant mass can be found to be diffused into the secondary porosity of the rock matrix when contaminants are in contact with bedrock. In these situations, the total mass estimated for a given system may be underestimated significantly if the contaminant mass is being derived solely from concentrations observed in groundwater. Finally, there are typically many different constituents in the subsurface environment that pose an additional electron demand other than the contaminant of concern (e.g., $NO_3^-$, $SO_4^{2-}$, etc.). Therefore, a holistic understanding of the contaminant mass and that of the various geochemical constituents present in a given system must be understood, to the degree practicable, to supply appropriate dosages of biostimulants.

The typical methods for determining the dosing requirement are (1) analysis of soil/rock and groundwater for the contaminant of concern plus a myriad of other organic and inorganic constituents that might be present, (2) analysis of the soil and groundwater for the contaminant of concern plus multiplication of a safety factor that is intended to estimate the additional demand from other constituents, (3) microcosm studies using both impacted aqueous and solid environmental media from the target area in an effort to determine empirical relationships between electron donor and electron acceptor.

The experience of the author suggests that an estimation of the total mass of contaminant in the system plus a safety factor along with microcosm studies provides the most robust calculation of dosing requirements. The microcosm studies will serve dual purposes: (1) they will serve as a "proof of concept" or determination that potential inhibitory constituents are absent, and (2) provides an understanding of the amount of electron donor/acceptor needed to drive the degradation/transformation reaction(s) to completion.

Once an estimate of the total mass of contaminant is known, a simple comparison of electron equivalents can be made to determine the corresponding stoichiometric amount of electron donor or acceptor needed to complete the reaction. For example, there are 12 milliequivalents per millimole of lactate and 6 milliequivalents per millimole of TCE. The simple calculation is to normalize the number of milliequivalents of lactate to those posed by the amount of TCE present to determine the mass of lactate required.

### Delivery techniques

As mentioned above, one of the most important aspects of successful *in situ* remediation is the ability to deliver the biostimulants (or any other material to aid in remediation) adequately to the area targeted for treatment. In the experience of the author, there have been misconceptions in the remediation community that injection of remediation enhancements or treatment solutions (in this case, biostimulants) occurs such that a 3-dimensional sphere is created by the delivery mechanisms. This is most certainly a false interpretation of what actually occurs in the subsurface. Rather, the injected fluid (or injectate) will follow the paths of increased permeability or hydraulic conductivity. That is to say, these fluids will take the path of least resistance. Since most soil matrices are not completely isotropic, the permeability will vary in 3-dimensions and it is therefore difficult to predict the travel paths of the injectate.

This difficulty in predicting the flow paths of the injectate is offset by the fact that most of these compounds will also be transported in the subsurface by advective, dispersive, and diffusive mass transport. Once the material is emplaced in the target area, the local microfauna will consume the material and in turn often produce a byproduct that serves as the actual electron donor (hydrogen for example). These byproducts will also be transported in 3-dimensions by the same mechanisms listed above such that spherical delivery is not necessary.

The method of delivery is also important to the success of the remediation project. There are several ways of emplacing the biostimulants into the target area. Installation of injection wells is used most often. These wells can simply be analogous to common groundwater monitoring wells. That is, the injection wells are often 2-inch diameter (approximately 5 centimeter diameter) wells constructed of polyvinyl chloride, stainless steel, or other similar rigid materials. The screens for these wells are typically shorter than the screens for common groundwater monitoring wells (2–5 feet or approximately 0.6–1.5 meters in length). The wells are installed with sand packs surrounding the screens with the annuli grouted appropriately.

The strategy for installing injection wells typically revolves around the fact that most *in situ* projects simply do not rely on a single injection. More typically, a given *in situ* project requires multiple injections over a longer period of time (sometime on the order of years rather than months). Therefore, semi-permanent injection wells serves to facilitate the longer-term nature of these projects and to control overall project costs.

There are situations when a single injection is all that is necessary. This may occur when the contamination is at very low levels and only just above clean-up goals. A single injection can be a very low cost method of reducing the contamination just enough to reach clean-up goals rather than relying on groundwater extraction (or even natural attenuation) to attain goals over a longer time frame. In this case, direct injection through drill rods is a common technique. The permeability of the subsurface is the single most important consideration when deciding to use this technique. The annulus surrounding the drilling rod is not sealed in this case, so if the subsurface permeability in the target area is more than $1 \times 10^{-4}$ centimeters/second, then installation of grouted monitoring wells is likely preferable (as a general rule of thumb). In any case, the advancement of the drill rods should most likely be by direct push methods (using a Geoprobe® for instance) with the rods being advanced very slowly and carefully. If the rods are shifted or flexed much during drilling, then the annulus becomes more exposed thereby creating a path of least resistance for transport of the injectate. This creates what is called a "surface expression" of the injectate. When the injectate surfaces, it is nearly impossible to recover the integrity of the injection point and a new injection location must be identified. Also, there are typically depth limitations to direct push technologies, such that if the target area is much more than 50 feet (again as a general rule of thumb), injection well installation may be required.

The injector heads that are used to allow pressurized injection are often patented or proprietary and come in many different varieties. Because of potential patent issues, a schematic of a typical injection head is not provided herein. However, design of an appropriate injector apparatus is not complicated and simply must allow a sealed connection between the injectate conveyance piping and appropriate gauges (flow meter, pressure gauge, etc.).

Injection velocities or flow rates will be highly site specific and be predicated on the hydraulic conductivity at the site in question. However, flow rates of 1–2 gallons per minute (0.004–0.008 cubic meters/minute) are not atypical. The injection pressure is also highly site specific and will typically be just greater than the hydrostatic pressure at the injection depth below ground surface.

### Radius of reactive influence

The conceptual illustration mentioned above relating to the distribution of biostimulants through paths of least resistance rather than a purely spherical pattern raises one of the most important aspects of *in situ* remedy implementation: the reactive radius of influence. Several different design aspects rest on a reliable understanding of this factor including, but not necessarily limited to, injection well spacing and required volume of biostimulant. A field-scale pilot program is typically employed to determine this radius. The pilot program is often comprised of a single injection well and a series of monitoring wells installed at various linear distances from the injection well. Samples are collected from the monitoring wells over time following an injection into the injection well to determine the radius of influence of the application. The "reactive" part of this design input is brought about by measuring the byproduct of the biostimulant. In other words, hydrogen is intended to be generated if an organic acid is used as the biostimulant. Therefore, dissolved hydrogen should be measured in the pilot monitoring wells to determine (1) the radius of lactate distribution, and (2) whether or not dissolved hydrogen is also resulting at the same radius or a distance further out.

Finally, it is ill-advised to attempt pressurized injection at depths less than 6–8 feet below ground surface. If injection is attempted at more shallow depths than this, there is a real and likely chance that the injectate will simply find its way back to the surface. Once this surface expression occurs, this injection point will be deemed inoperable and a new location will be required.

### In Situ Application Techniques—Bioaugmentation

The information provided above assumes that an appropriate microbial community is present at concentrations that will result in the desired treatment. There are situations where the appropriate bacteria (e.g., *Dehalococcoides* sp.; *Dhc*) are either at very low concentrations or absent altogether. For instance, consider the case of a PCE plume in a historically aerobic aquifer. PCE is not known to degrade under aerobic conditions, and the anaerobic microbes that degrade PCE are obligate anaerobes (cannot function or in some cases survive in the presence of oxygen). Given the aerobic conditions, a suitable *Dhc* population is most likely absent. Although anaerobic conditions can be created using an appropriate electron donor and *Dhc* may present itself, the system will likely require bioaugmentation. That is, an exogenous source of *Dhc* must be delivered to the subsurface environment.

There are many different commercially available microbial consortia. These are typically proprietary and must be obtained through the appropriate vendor (Geosyntec/SiREM—KB-1; CB&I—SDC-9; etc.). It should be noted that each of these consortia were isolated from different sites with differing conditions. This may mean that the various consortia are different in their metabolic characteristics. For instance, one consortia may have been grown on lactate while another may have been grown on methanol as electron donors. Some consortia may only degrade a contaminant in the presence of another (e.g., PCE degradation in the presence of VC). Therefore, it is very important to know the characteristics of the consortia one wishes to apply in order to successfully implement the technology and the desired site (Cashwell et al. 2004).

The need to bioaugment should be assessed with both time and economy in mind. There are cases (as in the PCE example above) when given enough time and an electron donor, an appropriate microbial population may develop. However, suitable monitoring requires expenditure as well as continued injection of the electron donor, when the microbial population may not develop after significant effort and expense. There are some who believe that biostimulation is the only thing needed to render an *in situ* bioremediation project successful, while others believe that bioaugmentation is generally required for a successful project. This author believes that they are both important and often necessary in tandem to attain the desired results.

Bioaugmentation is most commonly applied to systems in which anaerobic degradation is required due to the type of contaminant present in the target area. PCE, for instance, is only degraded in the absence of oxygen. These systems will be enhanced by addition of *Dhc*. The typical goal of the application is to reach a microbial population of at least $1 \times 10^7$ Dhc/L (Lu et al. 2006). The simple method of determining the amount of *Dhc* that must be present in the culture to be injected is to estimate the pore volume of the target area and multiply it by $1 \times 10^7$ (Aziz et al. 2013).

Since this application is most often used for anaerobic systems, the microbial consortia used in the process will be highly sensitive to atmospheric oxygen. This must be considered when preparing to deliver the consortia to the subsurface. The consortia must be delivered in air tight containers under a blanket of inert gas such as argon or nitrogen. The subsurface environment must also be anaerobic with evidence of nitrate and sulfate reduction prior to injection of the consortia. The equipment used to introduce the culture to the subsurface must be also purged with inert gas to maintain an anaerobic environment throughout the entire application process (Aziz et al. 2013).

Some vendors (e.g., Geosyntec) have used the "donut approach" to distribute anaerobic consortia for purposes of bioaugmentation. This approach strives to blanket the microbial consortia in anaerobic water or media in the subsurface. The intent is to avoid aeration of the target zone by injecting the aerobic biostimulant with the anaerobic consortia. This is done by initially injecting a small portion of the biostimulants followed by a surge of anaerobic water. The anaerobic consortia is then injected followed by another surge of anaerobic water. Finally, the remaining portion of the biostimulants is added. This approach has seen some success, but the actual result of the layered injection in the subsurface is likely not as robust as it may seem due to inevitable mixing caused by potential differences in viscosities and preferential pathways of these various materials. However, it is quite difficult (if not impossible) to de-oxygenate the biostimulants given the typical volumes of these materials and approaches such as the "donut approach" become ever more reasonable.

pH and temperature can greatly affect the success of bioaugmentation efforts. For *Dhc*, pH should optimally range from 6–8.3 standard pH units while the temperature of the subsurface should typically be between 10 and 30°C. There are many other geochemical considerations that should be considered when considering this application. The reader is encouraged to consult (Aziz et al. 2013) for more information.

## Conclusions

Due to the widespread use of chlorinated ethenes as industrial solvents, significant contamination has been found in many locations throughout the world. Due to their toxicity, cleanup of soil and groundwater contamination is needed in order to mitigate risks of exposure. In order to meet this need, *in situ* bioremediation techniques have been developed because they offer many advantages including completely mineralizing or transforming contaminants to harmless products. This book chapter examined the biodegradation and bioremediation of these compounds and provided practical insights in the design of treatment systems for their removal.

## Acknowledgements

The authors would like to acknowledge the Portuguese Science and Technology Foundation (FCT) under the Ciência 2008 program.

## References

Alvarez-Cohen, L. and G.E. Speitel. 2001. Kinetics of aerobic cometabolism of chlorinated solvents. Biodegradation 12: 105–126.

Aziz, C.E., R.A. Wymore and R.J. Steffan. 2013. Bioaugmentation considerations. pp. 141–169. *In*: H.F. Stroo, A. Leeson and C.H. Ward (eds.). Bioaugmentation for Groundwater Remediation. Springer Science + Business Media, New York.

Becker, J.G. and D.L. Freedman. 1994. Use of cyanocobalamin to enhance anaerobic biodegradation of chloroform. Environ. Sci. Technol. 28: 1942–1949.

Bradley, P.M. and F.H. Chapelle. 1996. Anaerobic mineralization of vinyl chloride in Fe(III)-reducing, aquifer sediments. Environ. Sci. Technol. 30: 2084–2086.

Bradley, P.M. and F.H. Chapelle. 1997. Kinetics of DCE and VC mineralization under methanogenic and Fe(III)-reducing conditions. Environ. Sci. Technol. 31: 2692–2696.

Bradley, P.M., F.H. Chapelle and D.R. Lovley. 1998. Humic acids as electron acceptors for anaerobic microbial oxidation of vinyl chloride and dichloroethene. Appl. Environ. Microbiol. 64: 3102–3105.

Cashwell, J.M., R. Marotte, S. Willis, J. Clarke, D. McLaughlin and D.L. Freedman. 2004. Evaluation of cultures for bioaugmentation of a PCE plume. Proceedings of the Battelle Fourth International Conference on Remediation of Chlorinated and Recalcitrant Compounds. Battelle Press, Columbus, OH, Monterey, CA.

Chang, Y.C., M. Hatsu, K. Jung, Y.S. Yoo and K. Takashima. 2000. Isolation and characterization of a tetrachloroethylene dechlorinating bacterium, *Clostridium bifermentans* DPH-1. J. Biosci. Bioeng. 89: 489–491.

Chuang, A.S. and T.E. Mattes. 2007. Identification of polypeptides expressed in response to vinyl chloride, ethene, and epoxyethane in *Nocardioides* sp. Strain JS614 by using peptide mass fingerprinting. Appl. Environ. Microbiol. 73: 4368–4372.

Coleman, N.V. and J.C. Spain. 2003a. Distribution of the coenzyme M pathway of epoxide metabolism among ethene- and vinyl chloride-degrading *Mycobacterium* strains. Appl. Environ. Microbiol. 69: 6041–6046.

Coleman, N.V. and J.C. Spain. 2003b. Epoxyalkane: coenzyme M transferase in the ethene and vinyl chloride biodegradation pathways of *Mycobacterium* Strain JS60. J. Bacteriol. 185: 5536–5545.

Coleman, N.V., T.E. Mattes, J.M. Gossett and J.C. Spain. 2002a. Biodegradation of cis-dichloroethene as the sole carbon source by a b-proteobacterium. Appl. Environ. Microbiol. 68: 2726–2730.

Coleman, N.V., T.E. Mattes, J.M. Gossett and J.C. Spain. 2002b. Phylogenetic and kinetic diversity of aerobic vinyl chloride-assimilating bacteria from contaminated sites. Appl. Environ. Microbiol. 68: 6162–6171.

Cowell, M.A., T.C.G. Kibbey, J.B. Zimmerman and K.F. Hayes. 2000. Partitioning of ethyoxylated nonionic surfactants in water/NAPL systems: effects of surfactants and NAPL properties. Environ. Sci. Technol. 34: 1583–1588.

Cupples, A.M., A.M. Spormann and P.L. McCarty. 2003. Growth of a Dehalococcoides-like microorganism on vinyl chloride and cis-dichloroethene as electron acceptors as determined by competitive PCR. Appl. Environ. Microbiol. 69: 953–959.

Danko, A.S., M. Luo, C.E. Bagwell, R.L. Brigmon and D.L. Freedman. 2004. Involvement of linear plasmids in aerobic biodegradation of vinyl chloride. Appl. Environ. Microbiol. 70: 6092–6097.

Danko, A.S., C. Saski, J.P. Tompkins and D.L. Freedman. 2006. Involvement of Coenzyme M in aerobic biodegradation of vinyl chloride and ethene of *Pseudomonas* strain AJ and *Ochrobactrum* strain TD. Appl. Environ. Microbiol. 72: 3756–3758.

Darlington, R., L. Lehmicke, R.G. Andrachek and D.L. Freedman. 2008. Biotic and abiotic anaerobic transformations of trichloroethene and cis-1,2-dichloroethene in fractured sandstone. Environ. Sci. Technol. 42: 4323–4330.

Department of Defense, U.S. 2008. Treatment of RDX and/or HMX Using Mulch Biowalls. Environmental Security Technology Certification Program.

Elango, V.K., A.S. Liggenstoffer and B.Z. Fathepure. 2006. Biodegradation of vinyl chloride and cis-dichloroethene by a *Ralstonia* sp. strain TRW-1. Appl. Microbiol. Biotechnol. 72: 1270–1275.

EPA (Environmental Protection Agengy, U.S.A.). 2004. *In situ* Thermal Treatment of Chlorinated Solvents: Fundamentals and Field Applications. p. 145.

Fathepure, B.Z., J.P. Nengu and S.A. Boyd. 1987. Anaerobic bacteria that dechlorinate perchloroethene. Appl. Environ. Microbiol. 53: 2671–2674.

Frascari, D., G. Zanaroli and A.S. Danko. 2015. *In situ* aerobic cometabolism of chlorinated solvents: A review. J. Hazard. Mater. 283: 382–399.

Freedman, D.L. and J.M. Gossett. 1989. Biological reductive dechlorination of tetrachloroethylene and trichloroethylene to ethylene under methanogenic conditions. Appl. Environ. Microbiol. 55: 2144–2151.

Giddings, C.G.S., J.C. Spain and J.M. Gossett. 2010. Microcosm assessment of a DNA probe applied to aerobic degradation of cis-1,2-Dichloroethene by *Polaromonas* sp. strain JS666. Ground Water Monit. R. 30: 97–105.

Gossett, J.M. 2010. Sustained aerobic oxidation of vinyl chloride at low oxygen concentrations. Environ. Sci. Technol. 44: 1405–1411.

Griffin, B.M., J.M. Tiedje and F.E. Löffler. 2004. Anaerobic microbial reductive dechlorination of tetrachloroethene to predominately trans-1,2-dichloroethene. Environ. Sci. Technol. 38: 4300–4303.

He, J., K.M. Ritalahti, M.R. Aiello and F.E. Löffler. 2003a. Complete detoxification of vinyl chloride by an anaerobic enrichment culture and identification of the reductively dechlorinating population as a *Dehalococcoides* species. Appl. Environ. Microbiol. 69: 996–1003.

He, J., K.M. Ritalahti, K.L. Yang, S.S. Koeningsberg and F.E. Löffler. 2003b. Detoxification of vinyl chloride to ethene coupled to growth of an anaerobic bacterium. Nature 424: 62–65.

IARC (International Agency for Research on Cancer). 1995. Dry-Cleaning, Some Chlorinated Solvents and Other Industrial Chemicals. World Health Organization IARC Monographs on the Evaluation of Carcinogenic Risks to Humans. Vol. 63.

Jennings, L.K., M.M.G. Chartrand, G. Lacrampe-Couloume, B.S. Lollar, J.C. Spain and J.M. Gossett. 2009. Proteomic and transcriptomic analyses reveal genes upregulated by cis-dichloroethene in *Polaromonas* sp. Strain JS666. Appl. Environ. Microbiol. 75: 3733–3744.

Katsenovich, Y., Z. Ozturk, M. Allen and G. Wein. 2007. Evaluation of compost materials for TCE biodegradation in shallow groundwater. pp. 1–10. *In*: K.C. Kabbes (ed.). World Environmental and Water Resources Congress Tampa, Florida.

Löffler, F.E., J. Yan, K.M. Ritalahti, L. Adrian, E.A. Edwards, K.T. Konstantinidis, J.A. Muller, H. Fullerton, S.H. Zinder and A.M. Spormann. 2013. *Dehalococcoides mccartyi* gen. nov., sp. nov., obligatelyorganohalide-respiring anaerobic bacteria relevant to halogen cycling and bioremediation, belong to a novel bacterial class, Dehalococcoidia classis nov., order Dehalococcoidales ord. nov. and family Dehalococcoidaceae fam. nov., within the phylum Chloroflexi. Int. J. Syst. Evol. Microbiol. 63: 625–635.

Lu, X., J.T. Wilson and D.H. Kampbell. 2006. Relationship between Dehalococcoides DNA in groundwater and rates of reductive dechlorination at field scale. Water Research 40: 3131–3140.

Madl, M.D. 2002. Aerobic Biodegradation of Ethene, cis-1,2-Dichloroethene and Vinyl Chloride in Sediments at the Savannah River Site Sanitary Landfill. Clemson University, Clemson, SC.

Marco-Urrea, E., I. Nijenhuis and L. Adrian. 2011. Transformation and carbon isotope fractionation of tetra- and trichloroethene to trans-dichloroethene by *Dehalococcoides* sp. Strain CBDB1. Environ. Sci. Technol. 45: 1555–1562.

Mattes, T.E., N.V. Coleman, J.C. Spain and J.M. Gossett. 2005. Physiological and molecular genetic analyses of vinyl chloride and ethene biodegradation in *Nocardioides* sp. strain JS614. Archives Microbiol. 183: 95–106.

Mattes, T.E., A.K. Alexander, P.M. Richardson, A.C. Munk, C.S. Han, P. Stothard and N.V. Coleman. 2008. The Genome of *Polaromonas* sp. Strain JS666: Insights into the evolution of a hydrocarbon- and xenobiotic-degrading bacterium, and features of relevance to biotechnology. Appl. Environ. Microbiol. 74: 6405–6416.

Maymó-Gatell, X., Y.T. Chien, J.M. Gossett and S.H. Zinder. 1997. Isolation of a bacterium that reductively dechlorinates tetrachloroethene to ethene. Science 276: 1568–1571.

Nishino, S.F., K.A. Shin, J.M. Gossett and J.C. Spain. 2013. Cytochrome P450 initiates degradation of cis-dichloroethene by *Polaromonas* sp. JS666. Appl. Environ. Microbiol. 79: 2263–2272.

Pontius, F.W. 1996. An update of the federal regs. Journal AWWA 88: 36–45.

Shimotori, T. and W.A. Arnold. 2002. Henry's law constants of chlorinated ethylenes in aqueous alcohol solutions: measurements, estimation, and thermodynamic analysis. J. Chem. Eng. Data 47: 183–190.

Schmidt, K.R., S. Gaza, A. Voropaev, S. Erti and A. Tiehm. 2014. Aerobic biodegradation of trichloroethene without auxilliary substrates. Water Research 59: 112–118.

Sung, Y., K.M. Ritalahti, R.P. Apkarian and F.E. Löffler. 2006. Quantitative PCR confirms purity of strain GT, a novel trichloroethene-to-ethene-respiring Dehalococcoides isolate. Appl. Environ. Microbiol. 72: 1980–1987.

Terzenbach, D.P. and M. Blaut. 1994. Transformation of tetrachloroethylene to trichloroethylene by homoacetogenic bacteria. FEMS Microbiol. Lett. 123: 213–218.

Vainberg, S. and R.J. Steffan. 2009. Large-scale production of bacterial consortia for remediation of chlorinated solvent-contaminated groundwater. J. Ind. Microbiol. Biotechnol. 36: 1189–1197.

Verce, M.F., R.L. Ulrich and D.L. Freedman. 2000. Characterization of an isolate that uses vinyl chloride as a growth substrate under aerobic conditions. Appl. Environ. Microbiol. 66: 3535–3542.

Vogel, T.M. and P.L. McCarty. 1985. Biotransformation of tetrachloroethylene to trichloroethylene, dichloroethylene, vinyl chloride, and carbon dioxide under methanogenic conditions. Appl. Environ. Microbiol. 49: 1080–1083.

# CHAPTER 3

# Eco-Labelling of Petrol Stations
## A Successful Experience in Brazil

*Angelo R.O. Guerra*[1,*] and *Francisco de A.O. Fontes*[1]

## ABSTRACT

This chapter describes a unique and successful experience concerning the Eco-labelling of petrol stations in Brazil. There are 110 fuel filling stations in the city of Natal (northeast of Brazil). No other city has ever undergone a program requiring that each facility would need to meet the highest degree of engineering design, operational and management systems. This practice is unique in the sense that all plants, even the ancient existing operational facilities, were required to upgrade and implement the engineering requirements in full. In 2009, The Public Prosecutor Office activated the Eco-labelling Program with the support of some experts from the academy, local environmental agency, and members of the fire department. Three main actions were undertaken aiming to guarantee its success: first, "Safety Reports" were produced, meaning an information pack containing an in-depth description of each facility including photos. Secondly, all facilities faced a tightness (leak) test in order to ensure that no pipework or Underground Storage Tanks (UST) were offering any risk to the environment. Thirdly, in case there was an on-going leak, the recovering of environmental liabilities would start immediately. Further steps led to negotiations and agreements, establishing the so-called "Terms of Adjustment of Conduct (TACs)", working with law enforcement. This chapter is focused on optimum engineering requirements necessary to both prevent and contain any loss of product with a special view to the Brazilian environmental legislation. Finally, it provides guidance towards the appropriate technical standards that should be adopted.

---

[1] Campus Universitário S/N, UFRN CT-DEM CEP 59078-900, Natal-RN, Brazil.
  Email: franciscofontes@uol.com.br
* Corresponding author: aroncalli@uol.com.br

## Introduction

According to the latest statistics produced by the Brazilian Petroleum, Natural Gas, and Biofuels Agency (ANP), Brazil is endowed with an extensive network of over 38,890 filling stations nationwide. The Eco-Labelling of Petrol Stations program reported in this chapter is an ongoing action conducted in the Brazilian state of Rio Grande do Norte, addressing approximately 557 filling stations. This chapter, however only focuses on the city of Natal (state capital) where there are exactly 110 petrol filling stations, which are already engaged in the program (Dias 2012).

This action is completely new and unique in its approach. The basis for this argument is twofold: first, no other state or city worldwide has ever reported to put forward a program that would include the whole set of filling stations from the four corners of their boundary. Starting in 2009, this program differs by including not only new sites and those being refurbished, but also older existing sites. Even older petrol stations were obliged to implement retrospectively the engineering requirements in full. Second, it is an innovative model for fostering partnership between the public prosecutor office, environmental agency, local fire department, academy, and organizations representing workers or employers from the petrol field. This has given all mentioned actors the opportunity to fine-tune the program as well as to discover important and needed changes.

What was the authorities' motivation for starting this program? Approximately 75% of Natal city's residents rely on groundwater to meet their water supply needs. Whenever petrol leaks or escapes from an underground storage tank or pipelines, it can have serious consequences in terms of health, environmental, economic, and resource issues. Furthermore, once the groundwater is polluted, it is extremely costly to restore its former unpolluted condition (i.e., to remediate). It must be highlighted that the release of petroleum hydrocarbons also poses a risk of fire or explosion and presents health and environmental hazards. Finally, one must bear in mind that concentrations of contamination may be at levels of unacceptable cancer risk. For instance, benzene is an organic chemical compound that can be found in gasoline and is known as a human carcinogen. Finally, one must bear in mind that fuel leaks and spills may inevitably happen because they are intrinsic to the fuel retail activity. The fluid tends to flow downward through the soil toward the groundwater table causing pollution.

Petrol station entrepreneurs must not neglect the importance of a tightness (leak) test. Over time, most metallic underground storage tanks and pipelines can corrode, crack, and develop leaks. Due to high-cost clean-up operations, preference should always be given to risk avoidance rather than risk control. Recently, the USA Environmental Protection Agency (EPA) reported that, as of March 2014, over 517,000 releases from federally-regulated leaking underground storage tanks were confirmed. The figures for Natal city in Brazil is that out of the 110 filling stations, 18 leaking underground storage tanks were confirmed and almost all petrol stations (99%) failed to pass the tightness (leak) test. The median concentration of benzene found in groundwater contaminated by leaking petrol stations was around 2,000 ppm, which is well above the drinking water standard of 0.005 ppm.

Finally, important environmental management measures need to be adopted by petrol stations as an effective tool for obtaining quick results while avoiding any environmental damage (Braga et al. 2002; Santos 2005; Adlmaier and Sellitto 2007; Barata et al. 2007; Lorenzett and Rossato 2010; Marques et al. 2013). In the next sections, we intend to draw attention to environmental issues of fuel activity and measures taken to comply with current Brazilian legislation.

## Engineering Requirements for Ecological Petrol Stations

This section presents the major ecological components of a petrol station designed to prevent spillages, overfilling, and leakages (Dias 2012; Escrivao et al. 2011; FECOMBUSTÍVEIS 2011; Leite 2006; Lorenzett and Rossato 2010; Lorenzett et al. 2011; Marques et al. 2013; Santos 2005) (Fig. 1). Some components also prevent the occurrence of atmospheric emission. At the end of this section, some important actions aimed at reducing the risk of accidents are described.

**Fig. 1.** Environmentally suitable petrol station fitted with ecological components.
Source: Adapted from Santos (2005).

## Leakage Prevention Devices

The very first engineering requirement is that dispensers' suction lines must be fitted with appropriate nozzle and under pump-valves (called check-valves—see Fig. 2). When the pump is turned off, the check valve closes and holds the product in the piping until the dispenser is reactivated for a new fuel filling procedure. This valve prevents leaks by continuously keeping a negative pressure inside the suction pipe. If there is a catastrophic line failure (holes, etc.), the suction is broken and the fuel drains back into the underground tank. In this way, the fuel will never be released into the environment.

Historically, all pipework already installed in Brazilian petrol stations was made of steel and was laid in a concrete surrounding (Dias 2012). It is known that the corrosion of both steel-made pipes and tanks usually occurs because of electrochemical reactions with chemical constituents of the enclosing soils.

**Fig. 2.** The Check-valve prevents leakage between tank and pump.
Source: Public Prosecutor Office.

Therefore, depending on the ground conditions, the steel-made pipework is most likely to be affected by corrosion and has its structural integrity compromised. In order to prevent leakage, modern engineering standards require non-metallic pipework made of HDPE (High Density Polyethylene) with inner lining (see Fig. 3). These pipes prevent leakage because they are not vapor permeable and are not subject to corrosion. Other advantages of composite pipes are flexibility and a reduced need for joints. The experience in Brazil showed that the most significant points of leakage are related to badly assembled and loose joints. As far as pipelines are concerned and based on real data, the Eco-Labelling Program pointed out the huge amount of fuel (169,517 liters/year) that would reach the vulnerable aquifer of Natal city if all old rusted and leaky metallic pipes had been kept in use so far.

**Fig. 3.** Non-metallic pipeline made of HDPE (High Density Polyethylene).
Source: Public Prosecutor Office.

Also within the forecourt area, there are other important components that must be installed to protect against leaks. The containment chamber of supply units (also called dispensers sumps) is one of them. Figure 4 illustrates a dispenser equipped with an HDPE-made containment chamber that was fitted underneath the pump (green containment). This unit is responsible for withholding any fuel leak coming from the pump as well as from all above pipelines and fittings.

**Fig. 4.** Dispenser Sump made of HDPE (High Density Polyethylene).
Source: Public Prosecutor Office.

Similarly, it is necessary to install a containment chamber under the Diesel filter units. Figure 5 illustrates a sump installed underneath the diesel filter.

Although many researchers believe the most significant potential points of fuel leaks in petrol stations are the underground tanks, the authors of this chapter do not strongly share this opinion. Indeed, the experience reported in

**Fig. 5.** A sump installed underneath the diesel filter.
Source: Public Prosecutor Office.

this book has shown that the most important potential points of leakage are the corroded metallic pipework and some badly mounted joints. The collected real field data from Natal city give the authors every reason to put forward this argument. The amount of detected fuel leaks coming from pipelines was 15% higher than the leaks coming from compromised metallic underground tanks. If one considers only those petrol stations equipped with modern non-metallic systems, this number would even be higher.

However, no one should underestimate leakage from tanks. Older tanks are most likely to be single skinned and constructed from steel. Therefore, it is not rare to see the integrity of this type of tank easily compromised by either corrosion or damage. In order to comply with the Brazilian engineering regulation NBR13786 Class 3 (i.e., cities that use groundwater as main water supply), the underground tanks must be double skinned with interstitial space (also known in Brazil as Eco-tank or Jacketed-tank) and manufactured according to Brazilian standards described in NBR13785.

In a simplified way, these double-skinned tanks (see Fig. 6 and Fig. 7) are essentially a tank within a tank. They can be understood as a composition of two tanks grouped together in a single block. First, one smaller diameter metal tank made of carbon steel ASTM A-36 is inserted into a second glass-fiber-reinforced thermoset resin corrosion-resistant tank with a larger diameter. This last tank works as a secondary containment providing a barrier between the metal tank and the environment.

**Fig. 6.** Double-skinned tanks ready for installation.
Source: Public Prosecutor Office.

**Fig. 7.** Double-skinned underground tanks being installed.
Source: Public Prosecutor Office.

In Fig. 6, only the external part of the composition (fiber in orange color) can be seen. Clearly, the matching inner metal tank is not visible. Lubricating facilities also require a non-metallic tank such as the one (smallest) featured in the upper left corner.

Figure 7 illustrates two jacketed tanks during their installation process. Noteworthy is the fact that, despite these tanks being double-skinned, the upper manholes of the tanks are made of steel and no skin is available in this section. Therefore, they could be easily attacked by corrosion, for instance, immediately after a rainy season in which intrusion water is sometimes found within the manhole's sumps.

By embracing the previously mentioned idea that jacketed-tanks are fundamentally a tank within another, the space in between them, ranging from the inner primary metallic tank to the outer larger secondary non-metallic tank, is referred to as the interstice, annular space, or interstitial space. Indeed, the interstitial space of a double-skinned underground storage tank is actually the empty volume between the primary metallic (inner) wall and the secondary non-metallic (outer) wall of the tank itself. The interstitial space of the tanks needs to be monitored by electronic sensors so that in case of any change in its fluid level or vacuum it could mean a breach of the inner or outer wall of the tank. The electronic sensors are connected to a device that produces an audible alarm in case of the presence of fluid within the jacket, indicating leakage. The Brazilian engineering standards require retailers to install sensors aiming to monitor leaks in several strategic places: in the interstices of tanks, dispensers sumps, filter sumps, and manholes (CONAMA 2000; Santos 2005). Figure 8a illustrates an example of the electronic equipment for leaks and tank level monitoring.

Beyond electronic leak monitoring, retailers must also install automatic tank level gauges providing wet stock inventory to help detect any escape of petrol from underground storage tanks and pipelines. This equipment is essential to provide consistent and accurate monitoring of petrol delivered, stored, and dispensed at each site in order to detect leaks from the underground tanks and connected pipeline system. There are also Brazilian brands that

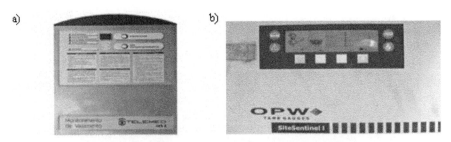

**Fig. 8.** Examples of equipment to monitor (a) leak detection and (b) tank levels.
Source: Public Prosecutor Office.

combine these two different functions (leak detection and wet stock inventory) in one single product. Figure 9 illustrates equipment combining leak detection and wet stock inventory functions.

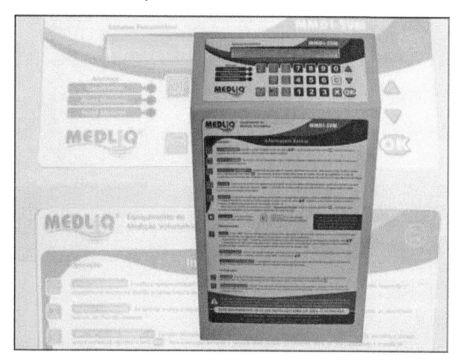

**Fig. 9.** Equipment combining leak detection and wet stock inventory functions. Source: Public Prosecutor Office.

## Spillage Prevention Devices

As far as petrol stations are concerned, control programs aimed at preventing leaks and accidental spillages must go hand-in-hand (CONAMA 2000; Santos 2005; Leite 2006; FECOMBUSTÍVEIS 2011; Marques et al. 2013). Therefore, spills also must not be allowed to seep into soil or water supplies. From a mechanical engineering standpoint, there are several equipment designed to help preventing fuel spills. NBR13786 provides the technical standards issued by the Brazilian Association of Technical Standards (ABNT) concerning this matter. First, the forecourt surface androad-tanker delivery stand are to be constructed of impermeable materials, free of gaps or cracks. Suitable materials include reinforced concrete or an approved equivalent. They must be provided with drainage channels to allow collection and direction of any spills to an oil water separator system (i.e., petrol interceptors).

In order to ensure the compliance with Brazilian environmental regulations, the wastewater from any vehicle washing systems must also be

**Fig. 10.** Forecourt impermeable surface and its drainage channel.
Source: Public Prosecutor Office.

**Fig. 11.** Tanker stand area with impermeable hard standing and its drainage channels.
Source: Public Prosecutor Office.

managed in a totally separate drainage system (i.e., separate interceptor) as the one used to drain the forecourt surface. Figure 12 illustrates an example of a vehicle washing system fitted with an impermeable surface.

**Fig. 12.** Washing system and offset filling fitted with impermeable surface and drainage channels. Source: Public Prosecutor Office.

The top of the tanks around the tank access lid must be fitted with liquid tight fiberglass containment chambers (i.e., manhole chambers). The installer must fix and seal these chambers sealed to the top of the tanks. They will take care of any leaks from pipe joints and flexible connectors (i.e., stainless steel annular convoluted hose). Figure 13 illustrates an example of a fiberglass containment chamber (blue color).

Another small equipment is the emergency breakaway device. They are designed to be installed on fuel dispensing hoses, and will separate when subjected to an unexpected pull force. They automatically stop the flow of fuel in order to avoid fuel spillage. Without having them fitted it would be a disaster if inadvertently a car moves while the dispensing nozzle is still attached to a car fuel tank. Figure 14 illustrates the breakaway device.

After installing all the above mentioned equipment, the polluted effluents will be drained and concentrated at the interceptor (also known as Oil/Water Separator). At this moment, as far as environmental issues are concerned, this equipment becomes the heart of the filling station. They usually are triple interceptors or equivalents and are capable of removing droplets. Firstly, the separator works as a wastewater treatment device for separating oil. The word oil in this particular context means liquid hydrocarbons that float

**Fig. 13.** Manhole chambers (blue containment) fitted on the tank.
Source: Public Prosecutor Office.

**Fig. 14.** Breakaway device to avoid spills at the dispenser area.
Source: Public Prosecutor Office.

on water such as diesel, petrol, and engine oil. However, its main task goes beyond this. After separating out any oil, they help the retailer by preventing the pollution from reaching the public, drains, sewers, or watercourses. In order to speed up the separation process, the separators are fitted with coalescence plates manufactured from both hydrophobic (water repelling) and oleophilic (oil attracting) materials. As the plates become coated with continuously agglomerating oil, the oil begins to form droplets faster. These droplets then coalesce or migrate upward. Oily waste must be transferred only to a registered waste carrier to be sent for recycling or disposal at a suitably licensed facility (CONAMA 2000; FECOMBUSTÍVEIS 2011; Lorenzett et al. 2011). Figure 15 illustrates an example of a coalescence oil/water separator.

**Fig. 15.** An example of a coalescence oil/water separator.
Source: Public Prosecutor Office.

## Overfill Prevention Devices

Most of the adverse events concerning both spills and overfills within a petrol station are related to routine activities. They often occur when tanks are filled and product is dispensed. Therefore, means should be provided to prevent accidental overfilling of a tank. The first device that helps to accomplish this task is the top-seal adaptor illustrated in Fig. 16. It also illustrates the catchment basins which are also called "spill containment manholes" or "spill buckets". Basically, a catchment basin is a bucket (blue) sealed around the fill pipe.

In this example, one can see the top-seal adaptor fitted with a y-shaped metallic central blockage that prevents the use of a stand-alone hose (i.e., without a quick coupler) which may cause overflows because of the

**Fig. 16.** Top-seal adaptor and spill container (blue color).
Source: Public Prosecutor Office.

clearances between the filling pipe and the hose itself. Instead, tight fill drop elbows should be used during delivery operations at the petrol station (Fig. 17). They are fitted between the tank filling tube and to the tank delivery hose. In between these last two parts are the top-seal adapters connecting delivery drop elbows to the fill pipe for fuel deliveries. It is important to mention here that these tight fill drop elbows also work as air pollution prevention devices.

It must be mentioned that overfill protection involves limiting the amount of fuel that can be delivered into a storage tank. This task is accomplished by is the overfill protection valve (Fig. 18) that should be installed in the drop tube of the tank and calibrated to prevent the tank from being filled in excess of 90% of its capacity.

The overfill protection valves slowly deliver petrol into the storage tank when the level in the storage tank approaches the design fill level and, therefore, prevent overtopping. They should be positioned to stop the petrol flow before the float vent valve (Fig. 19) operates. Float vent valves are also known as "float ball valves". If the fuel level becomes too high, the ball floats up and blocks the vent opening. Assuming the vent is blocked, very little fuel can flow into the tank and this will reduce the risk of overfilling. It is important to mention that for this float ball to work, the delivery hose must be tightly clamped to the fill pipe. Otherwise, fuel will back up the fill pipe and spill out of the fill opening.

**Fig. 17.** Tight fill drop elbows.
Source: Public Prosecutor Office.

**Fig. 18.** Overfill protection valve.
Source: Public Prosecutor Office.

**Fig. 19.** Float ball valve.
Source: Public Prosecutor Office.

## Air Pollution Prevention Devices

Beyond the previously mentioned tight fill drop elbows devices, in order to comply with environment regulations, all tank vapor vents should be equipped with pressure and vacuum vent valves. These vent valves usually operate in normally closed position by preventing vapor discharge through the vent pipe into the environment. When there is a variation in the inner pressure of the storage tank generated by the fill or delivery operation, it releases the

**Fig. 20.** Pressure and vacuum vent valves and vent collision barriers.
Source: Public Prosecutor Office.

vent pipe. These vents should terminate in open air in such a position that flammable vapors will not accumulate or travel to unsafe places. Therefore, they must extend to a height above the weather deck and must terminate at a standard distance from any living or working space, ventilator inlet, or source of ignition. Furthermore, suitable collision barriers resistant to damage from the impact of a motor vehicle should protect the venting system.

## Fire and Explosion Prevention Devices

Most activities involving petrol are potentially hazardous because the vapors given off by the liquid are highly flammable. Therefore, managing and controlling the risks of fire and explosion arising from the storage and dispensing of petrol is of the utmost importance (FECOMBUSTÍVEIS 2011; Dias 2012; Marques et al. 2013). Petrol is a volatile liquid, which gives off vapor flammable at very low temperatures and presents fire, explosion, health, and environmental hazards. When mixed with air in certain proportions the vapor forms a highly flammable atmosphere, which can burn or explode if ignited. A mixture containing about 1%–8% of petrol vapor in air is flammable.

The first safety rule is that any electrical equipment should be excluded from hazardous areas (Dias 2012). Where this is not possible, for example, electrical components in petrol dispensers, they must be constructed or protected to prevent danger (Fig. 21). The Brazilian Association of Technical Standards (ABNT) imposes a requirement to classify areas where explosive

**Fig. 21.** Conduit-sealing fittings above a dispenser sump.
Source: Public Prosecutor Office.

atmospheres may occur into zones based on their likelihood and persistence. These areas already classified into zones must be protected from sources of ignition by selecting adequate equipment and protective systems. For instance, fuel dispensers require conduit-sealing fittings like the one shown in Fig. 21 because they are designed to contain any explosion that occurs by preventing the passage of gases, vapors, or flames from one portion of a conduit system to another.

Figure 22 also illustrates an explosion-proof lamp fitted at a compressed natural gas station. Explosion-proof and flame-proof enclosures are capable of withstanding internal explosions and prevent ignition of the specified gases or vapors surrounding the electrical equipment.

**Fig. 22.** An explosion-proof lamp fitted at a CNG station.
Source: Public Prosecutor Office.

The most common fire and explosion hazards are associated with the unloading of road tankers. Although the receiving tanks at petrol stations are assumed to be providing the connection to a true earth ground, this is not always true. Electrostatic charge may accumulate on the tank or chassis of the road tank truck. This creates a risk of an electrostatic discharge igniting the explosive atmosphere nearby the fill drop pipe of the tank. One must bear in mind that an electrostatic discharge is always a spark. A true earth ground point should be installed in a distance greater than 3 m from the tank fill points in compliance with Brazilian engineering regulations. Figure 23 illustrates a ground point fitted away from an off-set fill pipe of the tanks.

Finally, fire extinguishers and a lightning protection system should be installed. The latter will provide means by which this discharge may enter or leave the earth without passing through and damaging non-conducting parts of a structure, such as those made of wood, brick, or concrete. Figure 24 illustrates an example of a lightning protection system.

**Fig. 23.** A ground point fitted away from an off-set fill pipe of the tanks.
Source: Public Prosecutor Office.

**Fig. 24.** An example of a lightning protection system.
Source: Public Prosecutor Office.

## Concluding Remarks

It is evident that the Brazilian Environmental Agency is getting more involved and has decided to put into practice a very stringent legislation. The agency started issuing both CONAMA Resolution 237/97 and CONAMA Resolution 273/2000 as guidance for managers of petrol filling stations to enable them to comply with the relevant environmental regulations. The success of this program has required efforts by all actors from the fuel sector and society in general. In Natal city (state of Rio Grande do Norte-Brazil), the work of the public prosecutor office was decisive to implement the Eco-Labelling of Petrol Stations Program. The importance of partnerships including the Federal University of Rio Grande Norte, through specific projects of the Departments of Mechanical and Chemical Engineering, must be highlighted here. This joint action has demonstrated to be a feasible alternative solution to ensure effective implementation of environmental regulations and its maintainability. It worked extremely well up until now and has been applied to all 110 petrol stations in Natal. Finally, due to its success, the team has decided to expand the Eco-Labelling Program to the countryside of the state and it is foreseen to reach all 600 filling stations by 2017.

## References

Adlmaier, D. and M.A. Sellitto. 2007. Embalagens retornáveis para transporte de bens manufaturados: um estudo de caso em logística reversa. Prod. 17: 395–406.

Barata, M.M.L., D.C. Kligerman and C.A. Minayo-Gomez. 2007. Gestão Ambiental no Setor Público: uma questão de relevância social e econômica. Ciência Saúde Coletiva 12: 165–170.

Braga, B., I. Hespanhol, J.G.L. Conejo, M.T.L. Barros, M. Spencer, M. Porto, N. Nucci, N. Juliano and S. Eiger. 2002. Introdução à Engenharia Ambiental. Prentice Hall, São Paulo, Brazil.

CONAMA. 2000. Resoluções: resolução n° 273 de 29 de novembro de 2000 (Available at: http://www.mma.gov.br/port/conama/res/res00/res27300.html, accessed on September 25, 2010).

Dias, G.M. 2012. Adequação Ambiental dos postos de combustíveis de Natal e recuperação da área degradada. Ministério Público do Estado do Rio Grande do Norte, Natal. pp. 352. ISBN: 978-85-60809-04-2.

Escrivao, G., M.S. Nagano and E. Escrivao Filho. 2011. A gestão do conhecimento na educação ambiental. Perspect. ciênc. inf. 16: 92–110.

FECOMBUSTÍVEIS. 2011. Meio Ambiente. Rio de Janeiro (Available at: http://www.fecombustiveis.org.br/meio-ambiente.html, accessed on April 15, 2011).

Leite, P.R. 2006. Logística Reversa: meio ambiente e competitividade. Pearson Prentice Hall, São Paulo, Brazil.

Lorenzett, D.B. and M.V. Rossato. 2010. A gestão de resíduos em postos de abastecimento de combustíveis. Revista Gestão Industrial 6: 110–125.

Lorenzett, D.B., M. Neuhaus, M.V. Rossato and L.P. Godoy. 2011. Gestão de recursos hídricos em postos de combustíveis. Diálogos & Ciência 9: 1–11.

Marques, R.D.S., F. Fedeli, R.G. Corrêa and D.V. Cristo. 2013. Postos de serviços Orientações para o controle ambiental. Série Gestão Ambiental 7, Rio de Janeiro, Brazil.

Santos, R.J.Sh. dos. 2005. A gestão ambiental em posto revendedor de combustíveis como instrumento de prevenção de passivos ambientais. Dissertação (Universidade Federal Fluminense), Niterói, Brazil. pp. 217.

# Biodegradation of Pyrethroid Pesticides

*Idalina Bragança,*[1,a] *Valentina F. Domingues,*[1,*]
*Paulo C. Lemos*[2] **and** *Cristina Delerue-Matos*[1,b]

## ABSTRACT

Pyrethroids are widely used as insecticides in agriculture, veterinary, and domestic applications. Their increasing use has become an environmental concern. Pyrethroid pesticides (SPs) residues have been frequently detected in soils and thus in agricultural samples. Some authors considered the microbial breakdown of insecticides the most important catabolic reaction in soil and can be crucial for development of pesticide decontamination. The available biodegradation reports showed that the majority of the studies involving the biodegradation of pyrethroid pesticides are done by pure cultures, considering isolated organisms (Bacteria and Fungi) capable of proficiently degrading different SPs even when present as the sole carbon source. Several pyrethroid-hydrolyzing enzymes have consistently been purified and characterized from various resources including metagenomes and organisms. This chapter shows that microorganisms' strains hold potential to be used in bioremediation of pyrethroid contaminated soils.

[1] REQUIMTE/LAQV, Instituto Superior de Engenharia do Porto, Instituto Politécnico do Porto, Rua Dr. António Bernardino de Almeida 431, 4200-072 Porto, Portugal.
[a] Email: linab_20@hotmail.com
[b] Email: cmm@isep.ipp.pt
[2] REQUIMTE/CQFB, FCT/Universidade Nova de Lisboa, Caparica, Portugal.
[*] Corresponding author: vfd@isep.ipp.pt

## Introduction

The use of pesticides is important and essential to protect and facilitate agricultural productivity, since pests are largely responsible for losses incurred during the production of food. Pests can be insects, mice, fungi, microorganisms, and undesirable plants (van der Hoff and van Zoonen 1999). Classification of pesticides can be made according to the Environmental Protection Agency (EPA) regarding the type of pest targeted (e.g., insecticides, fungicides, algaecide, herbicides, and nematicides) but they can also be classified regarding their chemical properties namely organophosphate, organochlorine, carbamate, and pyrethroid pesticides. The World Health Organization (WHO) recommended the classification of pesticides by hazard, that distinguishes between the more and the less hazardous forms of each pesticide, being based on the toxicity of the compound. Pyrethroid pesticides are synthetic insecticides derived from pyrethrins, natural compounds present in the pyrethrum extract from *Tanacetum cinerariae folium* (Kaneko 2010). The main active constituents of the extract are pyrethrin I and pyrethrin II with smaller amounts of the related cinerins and jasmolins. As can be seen in Fig. 1, the structural difference between the pyrethrum extract constituents is that pyrethrin I, cinerin I, and jasmolin I have a monocarboxylic acid (ester of chrysanthemic acid) while pyrethrin II, cinerin II, and jasmolin II have a dicarboxylic acid (ester of pyrethric acid) (Gosselin et al. 1984; Wakeling et al. 2012).

Synthetic pyrethroids were developed to preserve the insecticidal activity of pyrethrins and to enhance physical and chemical properties, such as increased stability in light (Gosselin et al. 1984). Pyrethroids are distinguished by three general characteristics: extreme hydrophobicity, rich stereochemistry, and broad-spectrum high-level insecticidal activity.

Pyrethroids are widely used as insecticides in agriculture, veterinary, and domestic applications (Albaseer et al. 2011; Bronshtein et al. 2012). The wide use of such pesticides in recent years is due to their advantages compared with other pesticides, such as its selectivity, easy degradation in the environment, and low acute toxicity to mammals (Albaseer et al. 2011; Bronshtein et al. 2012).

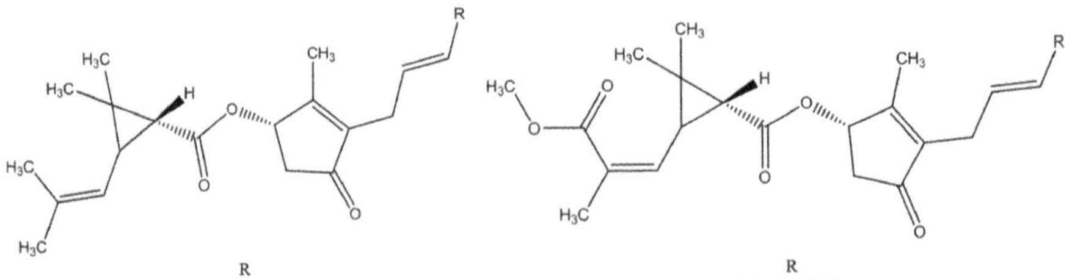

| | R |
|---|---|
| Pyrethrin I | CHCH₂ |
| Cinerin I | CH₃ |
| Jasmolin I | CH₂CH₃ |

| | R |
|---|---|
| Pyrethrin II | CHCH₂ |
| Cinerin II | CH₃ |
| Jasmolin II | CH₂CH₃ |

**Fig. 1.** Molecular structure of the six constituents of pyrethrum extract.

SPs unlike the naturally occurring pyrethrins are stable and persist longer in the environment and capture greater biological activity (Temple and Smith 1996). Pyrethroids may be classified by the presence or absence of a cyano group as type I if they are non-cyanopyrethroids or type II if they have the cyanogroup. Permethrin is an example of a type I pyrethroid while cypermethrin is a type II pyrethroid (Fig. 2) (Mullaley and Taylor 1994; Kurihara et al. 1997).

Worldwide pesticides consumption greatly increased following the increase of population and food production. Due to their extensive use, pyrethroid pesticides residues have been frequently detected in soils, sediments (Weston et al. 2013), and groundwater (Gonçalves et al. 2007).

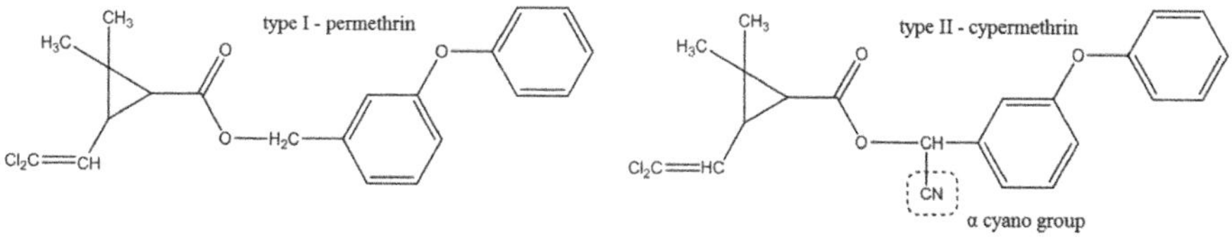

**Fig. 2.** Molecular structure of type I (permethrin) and type II (cypermethrin) synthetic pyrethroid pesticides.

Pyrethroid detections in environmental samples are commonly performed by gas chromatography (GC) coupled with the electron-capture detector (ECD) (Domingues et al. 2009) or mass spectrometry (MS), although there are other options based on tandem MS (MS$^2$) or comprehensive two-dimensional GC (GC × GC) coupled with time-of-flight MS (ToF-MS) (Feo et al. 2010). Contamination of agricultural samples (Columé et al. 2001) like fruits (Iñigo-Nuñez et al. 2010), vegetables (Akoto et al. 2013), and tea leaves (Nakamura et al. 1993) may result from preceding treatments in the soil and cross-contamination. The application of pyrethroids for veterinary purpose can be a source of contamination in animal derived products such as milk. Deltamethrin and cypermethrin were detected in residual levels in milk and lactea drink samples, which can be related to their wide use in dairy farming to kill ticks (Goulart et al. 2008).

The use and abuse of pyrethroid pesticides becomes a problem to the environment and thus to human health (Zhang et al. 2011b). Pyrethroids are recognized to contribute to sediment toxicity when found in concentrations which exceed the level expected to cause toxicity. For these reasons, remediation of pyrethroids contaminated soils is paramount to avoid public health hazards. Degradation of pyrethroid insecticides was observed to be more rapid with microorganisms versus sterilized soils, indicating that biological processes do contribute to breakdown in soil. Microbial degradation of SP appears to be a significant breakdown route of such pesticides (Palmquist et al. 2012). Microorganisms, as bacteria and fungi, are known to be primarily responsible for pesticide biodegradation in soil since they have several esterase enzymes capable of degrading pyrethroids (Singh 2002).

Biodegradation of pyrethroid insecticides by bacteria was observed to be more rapid than natural process. The half-lives ($T_{1/2}$) in the presence of the bacteria reported were significantly lower than the ones reported by Laskowski (2002). Pyrethroid pesticides are usually converted to 3-phenoxybenzaldehyde (3-PBAld) and 3-phenoxybenzoic acid (3-PBAcid) (Fig. 3).

In this chapter, some of the current knowledge on several aspects regarding pyrethroid pesticides, such as mode of action and impact on the environment and humans is discussed. An overview of the biodegradation of pyrethroids by different microorganisms is also given.

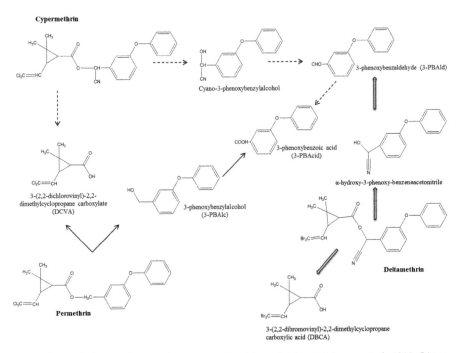

**Fig. 3.** Degradation pathway of some pyrethroid pesticides (Maloney et al. 1988; Liang et al. 2005; Tallur et al. 2008; Chen et al. 2011c).

## Mode of Action and Impact

Pyrethroids are considered neurotoxins since they affect the neuronal function. They act on neuronal receptors and/or ion channels (Zhou et al. 2011). These insecticides act primarily on sodium channels, which are essential to membrane excitability for generation and propagation of nerve action potentials (Dong 2007). However, sensitive calcium and chloride channels can also be an important target of pyrethroid action (Shafer and Meyer 2004). Pyrethroids type I and II share a similar mode of action, as they inhibit insects' nervous systems by keeping the sodium ion channels in nerve cell membranes open. Type I pyrethroids elicit depolarizing after potentials and repetitive discharges, whereas type II pyrethroids cause stimulus-dependent membrane depolarization leading to conduction block voltage-dependent membrane (Wakeling et al. 2012). Ray and Fry (2006) provided an overview of studies suggesting that the principal effects of pyrethroids were: neurotoxicity development, production of neuronal death, and action mediated via pyrethroid metabolite.

Regarding the persistence of SP in the environment, they became a concern due to their effect on non-target organisms. Although having a relatively low toxicity for mammals and birds, they are extremely toxic for fish and a variety of aquatic invertebrates (Khan 1983). Their increasing application generates concerns about their effects on aquatic invertebrates (Wang et al. 2011), long-term effects, and chronic toxicity in animal species and in humans (Albaseer et al. 2011). Human exposure to these compounds is high, with estimates indicating that 70% of the US population has detectable levels of pyrethroid metabolites in their urine (Ross 2011). There are studies that show that pyrethroid pesticides should be considered as hormone disruptors (Go et al. 1999), have carcinogenic potential (Shukla et al. 2002), and produce metabolites with endocrine activity (Tyler et al. 2000). Furthermore, studies in rats suggested that they can integrate the contribution of the environmental inductive component in the etiology of Parkinson's disease (Nasuti et al. 2007).

Some reports have seen effects of systemic poisoning by SPs. The type I pyrethroids produce tremors and reflex hyperexcitability (T intoxication syndrome). The type II variants produce symptoms that include choreoathetosis, salivation, and seizures (CS intoxication syndrome). Both types potentiate activation of the sympathetic nervous system (Ray and Forshaw 2000; Soderlund et al. 2002). Some pyrethroids clearly exhibit the historic classification symptoms of the T and CS syndromes while others do so less obviously and others, such as fenpropathrin, exhibited features of both groups (Weiner et al. 2009).

Overall, studies imply that each pyrethroid is unique in its ability to influence several cellular pathways. These findings suggest that pyrethroids can be toxic and hormone disruptors and their potential to affect humans and environmental ecosystems should be further investigated.

## Biodegradation

Bioremediation is a very promising technology with great potential for contaminated soils (Zouboulis and Moussas 2011). Bioremediation is increasing due to its efficiency and low costs associated compared with other technology (Vidali 2001). This process is mainly based on the utilization of soil microorganisms to degrade or detoxify pollutants. Microorganisms are primarily responsible for pesticide degradation in soil, with the microbial breakdown of insecticides being considered the most important catabolic reaction taking place in soil. Microorganisms use pesticides present in the environment for normal metabolic processes as carbon or phosphorus sources or other sources of food and energy. This microbial biodegradation process can be crucial when performing pesticide decontamination of the environment (Singh 2002).

To evaluate biodegradation kinetics of the SP some authors used a first order model $C_t = C_0 e^{-kt}$, where k is the rate constant (day$^{-1}$), $C_0$ is the amount of compound at time zero, and $C_t$ is the amount of compound at time t (days). The degradation half-life ($T_{1/2}$) is the time in which the compound concentration is reduced by 50%. The $T_{1/2}$ was determined by $T_{1/2} = \ln 2/k$. Concerning the obtained $R^2$, the pyrethroids biodegradation fit well with this model (Chen et al. 2011a, 2011c, 2011d). Other authors also used a second-order kinetic model to describe the biodegradation (Chen et al. 2012b). Usually, the reported $T_{1/2}$ for SPs in water varies from 17 to 600 days. Pyrethroids are degradable in soils with half-lives ranging from 3 to 96 days aerobically, and 5 to 430 days an aerobically (Laskowski 2002). However, pyrethroid persistence under aerobic soil conditions (field conditions) is highly variable. The half-lives can vary from 11.5 days for cyfluthrin to 96.3 days for bifenthrin. Degradation of pyrethroid insecticides was observed to be more rapid in natural versus sterilized soils, indicating that biological processes do contribute to the breakdown of these contaminants (Palmquist et al. 2012).

## Bacteria

*Bacteria* have a large genetic diversity with the potential to be used as a resource for the development of bioremediation processes. Several studies investigated the biodegradation of pyrethroid pesticides by these microbes in a controlled laboratory environment (Table 1).

The majority of the studies involving the biodegradation of pyrethroid pesticides were done with pure bacterial cultures.

*Acidomonas* sp. was capable of using allethrin as the sole carbon source and degraded more than 70% of allethrin in minimal medium within three days. The intermediates 2-ethyl-1,3-dimethyl-cyclopent-2-ene-carboxylic acid and cyclopropanecarboxylic acid, 2,2-dimethyl-3-(2-methyl-1-propenyl) were the major metabolites of allethrin biodegradation (Paingankar et al. 2005).

Most studies reporting SPs biodegradation by bacteria were done with cypermethrin and its active stereoisomers. *Azoarcus indigens* HZ5 was able to degrade 90% of β-cypermethrin in a mineral salts medium after six days. The metabolites formed were identified as 3-phenoxybenzaldehyde and 3-phenoxybenzoic acid (Ma et al. 2013). Cypermethrin was also degraded by *Micrococcus* sp. CPN 1, as sole energy source, in mineral salt medium being completely degraded. For the latter organism, cypermethrin was broken down to 3-phenoxybenzoate followed by protocatechuate, phenol, and then complete mineralization was achieved (Tallur et al. 2008). *Pseudomonas*

**Table 1.** Pyrethroids pesticides degradation by several bacterial species.

| Pesticides/ metabolite | Bacteria | Metabolites formed | Degradation | Reference |
|---|---|---|---|---|
| Allethrin | *Acidomonas* sp. | Major metabolites: Allethrolone; chrysanthemic acid; 2-ethyl-1,3 dimethyl cyclopent-2-enecarboxylic acid; cyclopropanecarboxylic acid and 2,2-dimethyl-3-(2-methyl-1-propenyl) | Allethrin (16 mM) was reduced to 49 and 25% initial concentration after 2 and 3 days of biodegradation, respectively, in minimal medium | (Paingankar et al. 2005) |
| β-Cypermethrin | *Azoarcusindigens* HZ5 | 3-phenoxybenzaldehyde and 3-phenoxybenzoic acid | β-cypermethrin in medium was 90% degraded after 6 days | (Ma et al. 2013) |
| Cypermethrin | *Bacillus* sp. ISTDS2 | 3-phenoxy phenyl hydroxyacetonitrile, carboxylate, 3-(2, 2-dichloroethenyl)-2, 2-dimethylcyclopropanecarboxylate (DCVA) and 3-PBA | Complete degradation of cypermethrin in soil microcosm after 30 days | (Sundaram et al. 2013) |
|  | *Bacillus cereus* ZH-3 *Streptomyces aureus* HP-S-01 | 4-phenoxyphenyl-2,2-dimethyl-propiophenone; 3-phenoxybenzaldehyde and α-hydroxy-3-phenoxy-benzeneacetonitrile | $T_{1/2}$ in medium by *B. cereus* ZH-3: 1.4 days; *S. aureus* HP-S-01: 1.8 days and mixed co-culture: 0.5 days. Mixed cultures completely metabolized cypermethrin (50 mg L$^{-1}$) in medium within 3 days | (Sundaram et al. 2013) |
|  | *Micrococcus* sp. CPN 1 | 3-phenoxybenzoate; protocatechuate; phenol and then complete mineralization | Cypermethrin at 1 g L$^{-1}$ in medium as the sole energy source was completely degraded | (Tallur et al. 2008) |
| Cyfluthrin Cyhalothrin Fenpropathrin Deltamethrin Bifenthrin β-Cypermethrin | *Brevibacteriumaureum* DG-12 | Cyfluthrin degradation: 2,2,3,3-Tetramethyl-cyclopropanemethanol; 4-Fluoro-3-phenexy-benzoic acid; 3,5-Dimethoxy phenol and Phenol | $T_{1/2}$ in medium (days): cyfluthrin 1.3; cyhalothrin 1.0; fenpropathrin 1.5; deltamethrin 1.7; bifenthrin 1.9 and β-cypermethrin 1.9. Cyfluthrin, cyhalothrin, fenpropathrin, deltamethrin, bifenthrin, and | (Chen et al. 2013) |

*Table 1 contd....*

*Table 1 contd.*

| Pesticides/ metabolite | Bacteria | Metabolites formed | Degradation | Reference |
|---|---|---|---|---|
| β-Cypermethrin 3-PBAcid β-Cyfluthrin Fenpropathrin, Cyhalothrin Deltamethrin | *Ochrobactrum lupin* DG-S-01 | ------ | β-cypermethrin were degraded 87.4%, 89.1%, 82.6%, 80.9%, 80.1% and 78.3% in medium after 5 days<br><br>$T_{1/2}$ in medium (days): β-cypermethrin 1.9; β-cyfluthrin 2.3; fenpropathrin 2.7; cyhalothrin 6.5; deltamethrin 8.2. Degradation kinetics rates in medium within 5 days were: β-cypermethrin 90.4%, β-cyfluthrin 80.8%, fenpropathrin 74.4%, cyhalothrin 56.2%, and deltamethrin 43.0%. 3-PBAcid (25–50 mg L$^{-1}$) in medium was 90% degraded within 9 days. | (Chen et al. 2011a) |
| β-Cyfluthrin | *Pseudomonas stutzeri* S1 | α-cyano-4-fluoro-3-phenoxybenzyl-3(2,2-dichlorovinyl)-2,2-dimethylcyclopropane carboxylate; 4-fluoro-3-phenoxy-a-cyanobenzylalcohol and 3(2,2-dichlorovinyl)-2,2-dimethyl cyclopropanecarboxylic acid | 94% degradation of β-cyfluthrin in medium within 8 days | (Saikia et al. 2005) |
| Fenpropathrin, Cypermethrin Permethrin Cyhalothrin Deltamethrin Fenvalerate Bifenthrin | *Sphingobium* sp. JZ-2 | Fenpropathrin degradation: 3-phenoxybenzaldehyde; 3-phenoxybenzoate; protocatechuate and catechol | Complete degradation of fenpropathrin (50 mg L$^{-1}$) in medium after 4 days incubation. Permethrin and cypermethrin degraded slower than fenpropathrin but faster than fenvalerate, deltamethrin, and cyhalothrin. Bifenthrin (the most persistent) was only 25% degraded within 4 days | (Guo et al. 2009) |

| Pesticides | Microorganism | Degradation products | Results | Reference |
|---|---|---|---|---|
| Deltamethrin 3-PBAld Cyfluthrin Bifenthrin Fenvalerate Fenpropathrin Permethrin Cypermethrin | *Streptomyces aureus* HP-S-01 | Deltamethrin degradation: α-Hydroxy-3-phenoxy-benzeneacetonitrile; 3-Phenoxybenzaldehyde and 2-Hydroxy-4-methoxy benzophenone | Completely removed deltamethrin (50–300 mg L$^{-1}$) in medium within 7 days. Complete degradation of 3-phenoxybenzaldehyde at 50, 100, and 200 mg L$^{-1}$, a was observed after 6, 7, and 7 days of incubation, respectively. $T_{1/2}$ in medium (days): cypermethrin 1.57; cyfluthrin 0.87; fenpropathrin 1.32; bifenthrin 0.90; deltamethrin 0.80; fenvalerate 1.08 and permethrin 1.56 | (Chen et al. 2011c) |
| Fenvalerate 3-PBAcid Deltamethrin β-Cypermethrin β-Cyfluthrin Cyhalothrin | *Stenotrophomonas* sp. ZS-S-01 | Fenvalerate degradation: 3-PBAcid and then complete mineralization | $T_{1/2}$ in medium (days): Fenvalerate 1.2, deltamethrin 1.3, 3-PBA 1.6, β-cypermethrin 1.9, β-cyfluthrin 2.0 and cyhalothrin 4.0 Fenvalerate was 80% removed from soil within 5 days | (Chen et al. 2011d) |
| Cis-Bifenthrin Permethrin | *Stenotrophomonas acidaminiphila* (BF6, BF24, BF28); *Aeromonas sobria* (PM-1); *Erwinia carotovora* (PM-2); *Yersinia frederiksenii* (PM-5) | ———— | $T_{1/2}$ in medium (days): cis-bifenthrin: *S. acidaminiphila* BF6-8, *S. acidaminiphila* BF24-58, *S. acidaminiphila* BF28-12 $T_{1/2}$ in medium (days): cis-permethrin: *sobria*-1.4, *E. carotovora*-1.3, *Y. frederiksenii*-1.5 | (Liu et al. 2005) |
| Bifenthrin | *Stenotrophomonas acidaminiphila* (BF6, BF24, BF28) | ———— | $T_{1/2}$ (days) in aqueous phase 1.3 to 5.5 $T_{1/2}$ (days) in field sediment 14.3 to 19.4 | (Lee et al. 2004) |
| 3-PBAcid | *Sphingomonas* SP-SC-1 | Phenol; catechol and 2-phenoxyphen | Completely degraded within 1 day in medium | (Tang et al. 2013) |
| | *Bacillus* sp. DG-02 | 3-(2-methoxyphenoxy) benzoic acid, protocatechuate, phenol, and 3,4-dihydroxy phenol | $T_{1/2}$ in non-sterilized and sterilized soils inoculated with *Bacillus* sp. DG-02: 3.4 and 4.1 days, respectively | (Chen et al. 2012a) |

*aeruginosa* CH7 was not only able to degrade β-cypermethrin but also utilized it as the sole carbon and energy source for growth and biosurfactant production (Zhang et al. 2011a).

Aco-culture of *Bacillus cereus* ZH-3 and *Streptomyces aureus* HP-S-01 significantly enhanced the degradation of cypermethrin. In the mixed co-cultures, lower half-lives ($T_{1/2}$ = 0.5 days) of cypermethrin were observed, as compared to the ones of the pure culture ($T_{1/2}$ = 1.4–1.8 days) (Chen et al. 2012b).

A bacterial strain isolated from active sludge, *Brevibacterium aureum* DG-12, was found effective to degrade cyfluthrin (87.4%), cyhalothrin (89.1%), fenpropathrin (82.6%), deltamethrin (80.9%), bifenthrin (80.1%), and β-cypermethrin (78.3%) in medium after five days. This bacterium degraded cyfluthrin by cleavage of both the carboxylester linkage and diaryl bond to form 2,2,3,3-tetramethyl-cyclopropanemethanol, 4-fluoro-3-phenexy-benzoic acid, 3,5-dimethoxy phenol, and phenol (Chen et al. 2013).

*Ochrobactrum lupin* DG-S-01 efficiently utilizes β-cypermethrin, β-cyfluthrin, fenpropathrin, cyhalothrin, and deltamethrin as growth substrate. The degradation kinetics rates reached up to 90.4%, 80.8%, 74.4%, 56.2%, and 43.0% within five days, respectively (Chen et al. 2011a).

*Pseudomonas stutzeri* S1, an isolate from an enrichment culture, was capable of degrading 94% of β-cyfluthrin in mineral salts medium within eight days. Products formed during degradation were identified as α-cyano-4-fluoro-3-phenoxybenzyl-3(2,2-dichlorovinyl)-2,2-dimethylcyclopropane carboxylate, 4-fluoro-3-phenoxy-α-cyanobenzylalcohol and 3(2,2-dichlorovinyl)-2,2-dimethyl cyclopropanecarboxylic acid (Saikia et al. 2005).

*Sphingobium* sp. JZ-2 completely degraded fenpropathrin in mineral salts medium after four days of incubation. Permethrin and cypermethrin were degraded slower than fenpropathrin but faster than fenvalerate, deltamethrin, and cyhalothrin. Only 25% of bifenthrin, the most persistent one, was degraded within four days (Guo et al. 2009).

Deltamethrin in medium was completely removed within seven days with a half-life of 0.80 days by a newly isolated *Streptomyces aureus* HP-S-01. Deltamethrin was metabolized by hydrolysis and produced α-hydroxy-3-phenoxybenzeneacetonitrile and 3-phenoxybenzaldehyde. This strain was also found to be highly efficient in degrading cyfluthrin, bifenthrin, fenvalerate, fenpropathrin, permethrin, and cypermethrin with slightly lower half-lives, ranging from 0.87 to 1.57 days (Chen et al. 2011c).

Fenvalerate was 80% removed from soil within five days by a newly isolated bacterium from activated sludge *Stenotrophomonas* sp. ZS-S-01. Fenvalerate and its hydrolysis product 3-PBAcid were degraded in medium with half-lives from 2.3 to 4.9 days. These bacteria could also degrade and utilize as growth substrates deltamethrin, β-cypermethrin, β-cyfluthrin, and cyhalothrin (Chen et al. 2011d).

*Stenotrophomonas acidaminiphila* BF6, BF24, and BF28 were capable of degrading bifenthrin with a $T_{1/2}$ between 1.3 to 5.5 days in aqueous phase and 14.3 to 19.4 days in field sediment (Lee et al. 2004). Using cis-bifenthrin other authors observed that for the same strains, rapid degradations of both enantiomers (1R-cis-bifenthrin and 1S-cis-bifenthrin) were attained with BF6 and BF28, whereas limited degradation was observed for the BF24 strain (Liu et al. 2005).

Three bacterial species reported to rapidly degrade permethrin in medium were identified as *Aeromonas sobria*, *Erwinia carotovora*, and *Yersinia frederiksenii*. Cis-permethrin degradation half-lives for these different bacteria ranged from 1.3 to 1.5 days (Liu et al. 2005).

Rare strains were reported to degrade cypermethrin in soil samples. A *Bacillus* sp. strain (ISTDS2) was described to completely accomplish this task but only after 30 days (Sundaram et al. 2013). Cypermethrin degradation was also tested in soil by the addition of *Pseudomonas* sp. Cyp19. Degradation of cypermethrin in un-autoclaved and autoclaved soil was 97.5% and 95% respectively after 30 days of incubation (Malik et al. 2009).

Regarding metabolite toxicity and impact on the environment, the goal of soil decontamination is not only the pesticides' degradation but also to accomplish complete mineralization. Additionally, some researchers show that *Ochrobactrum lupin* DG-S-01 (Chen et al. 2011a) degraded 90% of 3-PBAcid within nine days and *Sphingomonas* SP-SC-1 (Tang et al. 2013) completely degraded 3-PBAcid to phenol; catechol and 2-phenoxyphen within one day, in medium. *Streptomyces aureus* HP-S-01 was reported to completely degrade, in medium, the other main metabolite of SPs, 3-PBAld, after 6–7 days of incubation (Chen et al. 2011c). Complete mineralization of 3-PBAcid was achieved in soil by *Stenotrophomonas* sp. ZS-S-01 after 10 days (Chen et al. 2011d). A proven advantage of bacterial strains for metabolite elimination in soil was ascertained when using *Bacillus* sp. DG-02 inoculated in soil. It was observed that the $T_{1/2}$ for 3-PBAcid was greatly reduced (3.4–4.1 days) when compared to the non-inoculated control soil (101.9–187.3 days) (Chen et al. 2012a).

The biodegradation functionality of the microorganisms was mostly tested in ideal conditions (in medium with optimized temperature and pH) before its subsequent application *in situ*. The degradation potential of these strains, some without any toxic by products formed, revealed their significance as a biological agent for the remediation of pyrethroid soil contaminated sites.

## Fungi

Fungi are important microorganisms that possess biochemical and ecological capacity to degradexenobiotic compounds (Harms et al. 2011). However the potential use for fungi in bioremediation of pyrethroids has not received the attention it deserves. To our knowledge, there are significantly less reports about bioremediation of pyrethroids by *Fungi* when compared with *Bacteria*.

Saikia and Gopal (2004) studied β-cyfluthrin degradation in Czapek dox medium by five fungi in pure culture (*Trichodermaviride* 5-2, *Trichoderma viride*

2211, *Aspergillus niger*, *Aspergillus terricola*, and *Phanerochaete chrysoporium*). The degradation half-lives for the different fungal strains were *T. viride* 5-2 7.07–14.14 days, *T. viride* 2211 10.66–46.20 days, *Phanerochaete chrysoporium* 18.73 days, *A. terricola* 38.50–77.07 days, and *A. niger* 57.75 days.

A fungus strain isolated from activated sludge was identified as a *Cladosporium* sp. HU and was found efficient for pyrethroid degradation in mineral salt medium. Fenvalerate, fenpropathrin, and β-cypermethrin were completely degraded after five days while deltamethrin, bifenthrin, and permethrin were degraded 94.6%, 92.1%, and 91.6% at the end of the experiment, respectively. The degradation half-lives of each substrate ranged from 0.99 to 1.54 days. The fungus hydrolyzed fenvalerate in the carboxylester linkage to produce α-hydroxy-3-phenoxy-benzeneacetonitrile and 3-PBAld and then degraded these two compounds (Chen et al. 2011b).

A novel yeast strain ZS-02, isolated from activated sludge and identified as *Candida pelliculosa*, was found highly effective in degrading bifenthrin over a wide range of temperatures and pH. Under the optimal conditions, the yeast completely metabolized bifenthrin within eight days (Chen et al. 2012c).

These described applications of fungi used as pyrethroid degrader make fungi suited for use in pyrethroid-contaminated environments.

## Enzymes

Enzymes can also catalyze unspecific reactions and are actually recognized as promiscuous biocatalysts capable of transforming a variety of substrates that share structural similarity with their primary substrate (Aharoni et al. 2005; Khersonsky and Tawfik 2010). Some pyrethroid esterases have consistently been purified and characterized from various resources including metagenomes and organisms.

A novel pyrethroid-hydrolyzing enzyme Sys410 was recently isolated from Tuban Basin soil through a metagenomic approach. This enzyme efficiently degraded cyhalothrin, cypermethrin, sumicidin, and deltamethrin under assay conditions for 15 min, exceeding 95% hydrolysis efficiency (Fan et al. 2012).

A capable pyrethroid-hydrolyzing carboxylesterase was purified from mouse (Stok et al. 2004) and human (Nishi et al. 2006) liver microsomes. The carboxylesterases are enzymes that catalyze the hydrolysis of a wide range of ester-containing endogenous and xenobiotic compounds.

Enzyme extracts from bacterial and fungal strains were also reported as pyrethroid degraders. Purified enzyme from the fungus *Aspergillus niger* ZD11 was described to have a specific role on pyrethroid pesticides. Its hydrolysis action towards trans-permethrin was fastest, while deltamethrin was the least readily attacked. Cis-permethrin was hydrolyzed at an approximately equal rate as trans-permethrin (Liang et al. 2005). Pyrethroid detoxification was also achieved with a cell-free microbial enzyme system referred to as a *Bacillus cereus* SM3 permethrinase. Researchers think that this esterase seems to be a carboxylesterase (Maloney et al. 1993). The pyrethroid-hydrolyzing esterase gene from *Klebsiella* sp. ZD112 was cloned and sequenced. There combinant gene encoding pyrethroid-hydrolyzing esterase (EstP) was purified and characterized. The purified EstP not only degraded many pyrethroid pesticides and the organophosphorus insecticide malathion, but also hydrolyzed F-nitrophenyl esters of various fatty acids, indicating that EstP is an esterase with broad substrate range. Trans- and cis-permethrin hydrolyzes indicate higher hydrolyzes efficiency than the carboxylesterases from resistant insects and mammals (Wu et al. 2006).

These pyrethroid-hydrolyzing enzymes could conceivably be developed to fulfill the requirements to enable its use *in situ* for detoxification of pyrethroids where they cause environmental problems.

## Conclusion

Pyrethroids' wide use as insecticides in agriculture, veterinary, and domestic applications raises environmental concerns. Pyrethroid pesticides residues have been frequently detected in soils and thus in agricultural samples. The microbial breakdown of insecticides is considered by some authors as the most important catabolic reaction in soil and can be crucial when developing processes for pesticide decontamination.

Biodegradation reports showed that isolated microorganisms and mixed cultures were capable of proficiently degrading different SPs even when present as the sole carbon source. Bacteria, fungi, and enzymes possess potential for their use in bioremediation of pyrethroid-contaminated soils.

## References

Aharoni, A., L. Gaidukov, O. Khersonsky, Q.G.S. Mc, C. Roodveldt and D.S. Tawfik. 2005. The 'evolvability' of promiscuous protein functions. Nat. Genet. 37: 73–76.

Akoto, O., H. Andoh, G. Darko, K. Eshun and P. Osei-Fosu. 2013. Health risk assessment of pesticides residue in maize and cowpea from Ejura, Ghana. Chemosphere 92: 67–73.

Albaseer, S.S., R.N. Rao, Y.V. Swamy and K. Mukkanti. 2011. Analytical artifacts, sample handling and preservation methods of environmental samples of synthetic pyrethroids. Trac-Trends Anal. Chem. 30: 1771–1780.

Bronshtein, A., J.C. Chuang, J.M. Van Emon and M. Altstein. 2012. Development of a multianalyteenzyme-linked immunosorbent assay for permethrin and Aroclors and its implementation for analysis of soil/sediment and house dust extracts. J. Agric. Food. Chem. 60: 4235–4242.

Cha, K.Y., M. Crimi, A. Urynowicz and R.C. Bordem. 2012. Kinetics of permanganate consumption by natural oxidant demand in aquifer solids. Environ. Eng. Sci. 29: 646–653.

Chen, S., M. Hu, J. Liu, G. Zhong, L. Yang, M. Rizwan-ul-Haq and H. Han. 2011a. Biodegradation of beta-cypermethrin and 3-phenoxybenzoic acid by a novel *Ochrobactrum lupin* DG-S-01. J. Hazard. Mater. 187: 433–440.

Chen, S., Q. Hu, M. Hu, J. Luo, Q. Weng and K. Lai. 2011b. Isolation and characterization of a fungus able to degrade pyrethroids and 3-phenoxybenzaldehyde. Bioresour. Technol. 102: 8110–8116.

Chen, S., K. Lai, Y. Li, M. Hu, Y. Zhang and Y. Zeng. 2011c. Biodegradation of deltamethrin and its hydrolysis product 3-phenoxybenzaldehyde by a newly isolated *Streptomyces aureus* strain HP-S-01. Appl. Microbiol. Biotechnol. 90: 1471–1483.

Chen, S., L. Yang, M. Hu and J. Liu. 2011d. Biodegradation of fenvalerate and 3-phenoxybenzoic acid by a novel *Stenotrophomonas* sp. strain ZS-S-01 and its use in bioremediation of contaminated soils. Appl. Microbiol. Biotechnol. 90: 755–767.

Chen, S., W. Hu, Y. Xiao, Y. Deng, J. Jia and M. Hu. 2012a. Degradation of 3-phenoxybenzoic acid by a *Bacillus* sp. PLoS One 7, e50456.

Chen, S., J. Luo, M. Hu, K. Lai, P. Geng and H. Huang. 2012b. Enhancement of cypermethrin degradation by a coculture of *Bacillus cereus* ZH-3 and *Streptomyces aureus* HP-S-01. Bioresour. Technol. 110: 97–104.

Chen, S., Y.H. Dong, C. Chang, Y. Deng, X.F. Zhang, G. Zhong, H. Song, M. Hu and L.H. Zhang. 2013. Characterization of a novel cyfluthrin-degrading bacterial strain *Brevibacterium aureum* and its biochemical degradation pathway. Bioresour. Technol. 132: 16–23.

Chen, S.H., J.J. Luo, M.Y. Hu, P. Geng and Y.B. Zhang. 2012c. Microbial detoxification of bifenthrin by a novel yeast and its potential for contaminated soils treatment. PLoS One 7.

Columé, A., S. Cárdenas, M. Gallego and M. Valcárcel. 2001. Selective enrichment of 17 pyrethroids from lyophilised agricultural samples. J. Chromatogr. A 912: 83–90.

Domingues, V., M. Cabral, A. Alves and C. Delerue-Matos. 2009. Use and reuse of SPE disks for the determination of pyrethroids in water by GC-ECD. Anal. Lett. 42: 706–726.

Dong, K. 2007. Insect sodium channels and insecticide resistance. Invert Neurosci. 7: 17–30.

EPA. 2012. About Pesticides. Retrieved in http://www.epa.gov/pesticides/about/types.htm.

Fan, X., X. Liu, R. Huang and Y. Liu. 2012. Identification and characterization of a novel thermostable pyrethroid-hydrolyzing enzyme isolated through metagenomic approach. Microb. Cell Fact. 11: 33.

Feo, M.L., E. Eljarrat and D. Barceló. 2010. Determination of pyrethroid insecticides in environmental samples. TrAC-Trend Anal. Chem. 29: 692–705.

Gayer, K.H. and L. Woontner. 1956. The solubility of ferrous hydroxide and ferric hydroxide in acidic and basic media at 25°. J. Phys. Chem. 60: 1569–1571.

Go, V., J. Garey, M.S. Wolff and B.G.T. Pogo. 1999. Estrogenic potential of certain pyrethroid compounds in the MCF-7 human breast carcinoma cell line. Environ. Health Perspect. 107: 173–177.

Gonçalves, C.M., J.C.G. Esteves da Silva and M.F. Alpendurada. 2007. Evaluation of the pesticide contamination of groundwater sampled over two years from a vulnerable zone in Portugal. J. Agric. Food. Chem. 55: 6227–6235.

Gosselin, R.E., R.P. Smith, H.C. Hodge and J.E. Braddock. 1984. Clinical Toxicology of Commercial Products. Williams & Wilkins, Baltimore.

Goulart, S.M., M.E.L.R. de Queiroz, A.A. Neves and J.H. de Queiroz. 2008. Low-temperature clean-up method for the determination of pyrethroids in milk using gas chromatography with electron capture detection. Talanta 75: 1320–1323.

Guo, P., B. Wang, B. Hang, L. Li, S.W. Ali, J. He and S. Li. 2009. Pyrethroid-degrading *Sphingobium* sp. JZ-2 and the purification and characterization of a novel pyrethroid hydrolase. Int. Biodeterior. Biodegrad. 63: 1107–1112.

Harms, H., D. Schlosser and L.Y. Wick. 2011. Untapped potential: exploiting fungi in bioremediation of hazardous chemicals. Nat. Rev. Microbiol. 9: 177–192.

Iñigo-Nuñez, S., M.A. Herreros, T. Encinas and A. Gonzalez-Bulnes. 2010. Estimated daily intake of pesticides and xenoestrogenic exposure by fruit consumption in the female population from a Mediterranean country (Spain). Food Control 21: 471–477.

Kaneko, H. 2010. Pyrethroid chemistry and metabolism. pp. 1635–1663. *In*: R. Krieger (ed.). Hayes' Handbook of Pesticide Toxicology (Third Edition). Academic Press, New York.

Khan, N.Y. 1983. An assessment of the hazard of synthetic pyrethroid insecticides to fish and fish habitat. pp. 437–450. *In*: S. Matsunaka, D.H. Hutson and S.D. Murphy (eds.). Mode of Action, Metabolism and Toxicology. Pergamon, New York.

Khersonsky, O. and D.S. Tawfik. 2010. Enzyme promiscuity: a mechanistic and evolutionary perspective. Annu. Rev. Biochem. 79: 471–505.

Kurihara, N., J. Miyamoto, G.D. Paulson, B. Zeeh, M.W. Skidmore, R.M. Hollingworth and H.A. Kuiper. 1997. Pesticides report. 37. Chirality in synthetic agrochemicals: Bioactivity and safety consideration. Pure & Appl. Chem. 69: 1335–1348.

Laskowski, D.A. 2002. Physical and chemical properties of pyrethroids. Rev. Environ. Contam. Toxicol. 174: 49–170.

Lee, S., J.K. Gan, J.S. Kim, J.N. Kabashima and D.E. Crowley. 2004. Microbial transformation of pyrethroid insecticides in aqueous and sediment phases. Environ. Toxicol. Chem. 23: 1–6.

Lewis, S., A. Lynch, L. Bachas, S. Hampson, L. Ormsbee and D. Bhattacharyya. 2009. Chelate-modified fenton reaction for the degradation of trichloroethylene in aqueous and two-phase systems. Environ. Eng. Sci. 26: 849–859.

Liang, W.Q., Z.Y. Wang, H. Li, P.C. Wu, J.M. Hu, N. Luo, L.X. Cao and Y.H. Liu. 2005. Purification and characterization of a novel pyrethroid hydrolase from *Aspergillus niger* ZD11. J. Agric. Food. Chem. 53: 7415–7420.

Liu, W.P., J.Y. Gan, S. Lee and I. Werner. 2005. Isomer selectivity in aquatic toxicity and biodegradation of bifenthrin and permethrin. Environ. Toxicol. Chem. 24: 1861–1866.

Ma, Y., L. Chen and J. Qiu. 2013. Biodegradation of beta-cypermethrin by a novel *Azoarcus indigens* strain HZ5. J. Environ. Sci. Health B48: 851–859.

Malik, D., M. Singh and P. Bhatia. 2009. Biodegradation of cypermethrin by a Pseudomonas Strain Cyp19 and its use in bioremediation of contaminated soil. Internet J. Microbiol. 6.

Maloney, S.E., A. Maule and A.R. Smith. 1988. Microbial transformation of the pyrethroid insecticides: permethrin, deltamethrin, fastac, fenvalerate, and fluvalinate. Appl. Environ. Microbiol. 54: 2874–2876.

Maloney, S.E., A. Maule and A.R. Smith. 1993. Purification and preliminary characterization of permethrinase from a pyrethroid-transforming strain of *Bacillus cereus*. Appl. Environ. Microbiol. 59: 2007–2013.

Mullaley, A. and R. Taylor. 1994. Conformational properties of pyrethroids. J. Comput. Aid. Mol. Des. 8: 135–152.

Nakamura, Y., Y. Tonogai, Y. Tsumura and Y. Ito. 1993. Determination of pyrethroid residues in vegetables, fruits, grains, beans, and green tea leaves: applications to pyrethroid residue monitoring studies. J. AOAC Int. 76: 1348–1361.

Nasuti, C., R. Gabbianelli, M.L. Falcioni, A. Di Stefano, P. Sozio and F. Cantalamessa. 2007. Dopaminergic system modulation, behavioral changes, and oxidative stress after neonatal administration of pyrethroids. Toxicology 229: 194–205.

Nishi, K., H. Huang, S.G. Kamita, I.H. Kim, C. Morisseau and B.D. Hammock. 2006. Characterization of pyrethroid hydrolysis by the human liver carboxylesterases hCE-1 and hCE-2. Arch. Biochem. Biophys. 445: 115–123.

Paingankar, M., M. Jain and D. Deobagkar. 2005. Biodegradation of allethrin, a pyrethroid insecticide, by an *Acidomonas* sp. Biotechnol. Lett. 27: 1909–1913.

Palmquist, K., J. Salatas and A. Fairbrother. 2012. Pyrethroid insecticides: use, environmental fate, and ecotoxicology. pp. 251–278. *In*: D.F. Perveen (ed.). Insecticides—Advances in Integrated Pest Management. InTech, Retrieved in http://www.intechopen.com/books/insecticides-advances-in-integrated-pest-management/pyrethroid-insecticides-use-environmental-fate-and-ecotoxicology.

Ray, D.E. and P.J. Forshaw. 2000. Pyrethroid insecticides: poisoning syndromes, synergies, and therapy. J. Toxicol. Clin. Toxicol. 38: 95–101.

Ray, D.E. and J.R. Fry. 2006. A reassessment of the neurotoxicity of pyrethroid insecticides. Pharmacol. Therapeut. 111: 174–193.

Ross, M.K. 2011. Pyrethroids. pp. 702–708. *In*: J.O. Nriagu (ed.). Encyclopedia of Environmental Health. Elsevier, Burlington.

Saikia, N. and M. Gopal. 2004. Biodegradation of beta-cyfluthrin by fungi. J. Agric. Food. Chem. 52: 1220–1223.

Saikia, N., S.K. Das, B.K.C. Patel, R. Niwas, A. Singh and M. Gopal. 2005. Biodegradation of beta-cyfluthrin by *Pseudomonas stutzeri* strain S1. Biodegradation 16: 581–589.

Shafer, T.J. and D.A. Meyer. 2004. Effects of pyrethroids on voltage-sensitive calcium channels: a critical evaluation of strengths, weaknesses, data needs, and relationship to assessment of cumulative neurotoxicity. Toxicol. Appl. Pharm. 196: 303–318.

Shukla, Y., A. Yadav and A. Arora. 2002. Carcinogenic and cocarcinogenic potential of cypermethrin on mouse skin. Cancer Lett. 182: 33–41.

Singh, D.K. 2002. Microbial degradation of insecticides: an assessment for its use in bioremediation. pp. 175–188. *In*: P. Singh Ved and D. Stapleton Raymond (eds.). Progress in Industrial Microbiology. Elsevier, Amsterdam, The Netherlands.

Soderlund, D.M., J.M. Clark, L.P. Sheets, L.S. Mullin, V.J. Piccirillo, D. Sargent, J.T Stevens and M.L. Weiner. 2002. Mechanisms of pyrethroid neurotoxicity: implications for cumulative risk assessment. Toxicology 171: 3–59.

Stok, J.E., H.Z. Huang, P.D. Jones, C.E. Wheelock, C. Morisseau and B.D. Hammock. 2004. Identification, expression, and purification of a pyrethroid-hydrolyzing carboxylesterase from mouse liver microsomes. J. Biol. Chem. 279: 29863–29869.

Sundaram, S., M.T. Das and I.S. Thakur. 2013. Biodegradation of cypermethrin by *Bacillus* sp. in soil microcosm and *in-vitro* toxicity evaluation on human cell line. Int. Biodeterior. Biodegrad. 77: 39–44.

Tallur, P.N., V.B. Megadi and H.Z. Ninnekar. 2008. Biodegradation of cypermethrin by *Micrococcus* sp. strain CPN 1. Biodegradation 19: 77–82.

Tang, J., K. Yao, S. Liu, D. Jia, Y. Chi, C. Zeng and S. Wu. 2013. Biodegradation of 3-phenoxybenzoic acid by a novel *Sphingomonas* sp. SC-1. Fresen. Environ. Bull. 22: 1564–1572.

Temple, W. and N.A. Smith. 1996. 20—Insecticides. pp. 541–550. *In*: J. Descotes (ed.). Human Toxicology. Elsevier Science B.V., Amsterdam.

Tyler, C.R., N. Beresford, M. van der Woning, J.P. Sumpter and K. Tchorpe. 2000. Metabolism and environmental degradation of pyrethroid insecticides produce compounds with endocrine activities. Environ. Toxicol. Chem. 19: 801–809.

van der Hoff, G.R. and P. van Zoonen. 1999. Trace analysis of pesticides by gas chromatography. J. Chromatogr. A 843: 301–322.

Vidali, M. 2001. Bioremediation. An overview. Pure Appl. Chem. 73: 1163–1172.

Wakeling, E.N., A.P. Neal and W.D. Atchison. 2012. Pyrethroids and their effects on ion channels. *In*: R.P. Soundararajan (ed.). Pesticides—Advances in Chemical and Botanical Pesticides. InTech, Retrieved in http://www.intechopen.com/books/pesticides-advances-in-chemical-and-botanical-pesticides/pyrethroids-and-their-effects-on-ion-channels.

Wang, B.Z., Y. Ma, W.Y. Zhou, J.W. Zheng, J.C. Zhu, J. He and S.P. Li. 2011. Biodegradation of synthetic pyrethroids by Ochrobactrumtritici strain pyd-1. World J. Microbiol. Biotechnol. 27: 2315–2324.

Weiner, M.L., M. Nemec, L. Sheets, D. Sargent and C. Breckenridge. 2009. Comparative functional observational battery study of twelve commercial pyrethroid insecticides in male rats following acute oral exposure. Neurotoxicology 30 Suppl. 1: S1–16.

Weston, D.P., Y. Ding, M. Zhang and M.J. Lydy. 2013. Identifying the cause of sediment toxicity in agricultural sediments: the role of pyrethroids and nine seldom-measured hydrophobic pesticides. Chemosphere 90: 958–964.

WHO. 2009. The WHO Recommended Classification of Pesticides by Hazard.

Wu, P.C., Y.H. Liu, Z.Y. Wang, X.Y. Zhang, H. Li, W.Q. Liang, N. Luo, J.M. Hu, J.Q. Lu, T.G. Luan and L.X. Cao. 2006. Molecular cloning, purification, and biochemical characterization of a novel pyrethroid-hydrolyzing esterase from *Klebsiella* sp. strain ZD112. J. Agric. Food. Chem. 54: 836–842.

Zhang, C., S. Wang and Y. Yan. 2011a. Isomerization and biodegradation of beta-cypermethrin by *Pseudomonas aeruginosa* CH7 with biosurfactant production. Bioresour. Technol. 102: 7139–7146.

Zhang, W.J., F.B. Jiang and J.F. Ou. 2011b. Global pesticide consumption and pollution: with China as a focus. Proc. Int. Acad. Ecol. Environ. Sci. 1: 125–144.

Zhou, T., W. Zhou, Q. Wang, P.L. Dai, F. Liu, Y.L. Zhang and J.H. Sun. 2011. Effects of pyrethroids on neuronal excitability of adult honeybees Apismellifera. Pestic. Biochem. Phy. 100: 35–40.

Zouboulis, A.I. and P.A. Moussas. 2011. Groundwater and soil pollution: bioremediation. pp. 1037–1044. *In*: J.O. Nriagu (ed.). Encyclopedia of Environmental Health. Elsevier, Burlington.

# CHAPTER 5

# *In Situ* Chemical Oxidation (ISCO)

*Aurora Santos[1],\* and Juana Mª Rosas[2]*

## ABSTRACT

Chemical or biological techniques can be applied for the complete or substantial destruction/degradation of the pollutants in soils. Biological techniques require, in many cases, high treatment times (if residual Non Aqueous Phase Liquids (NAPL), high concentration plumes, and non-biodegradable compounds are present). On the other hand, *In Situ* Chemical Oxidation (ISCO) has been proven to be successful in the abatement of many organic pollutants. Permanganate, Ozone, Catalyzed Hydrogen Peroxide (CHP, Fenton's Reagents), and Persulfate have been used as oxidants. The key factors in ISCO applications are related to the chemistry of the oxidants in the subsurface and the oxidant delivery in a complex hydrogeological system. The oxidation can go through a radical or non-radical mechanism, having hydroxyl and sulfate radicals the highest oxidant strength. Besides this, the desired pollutant abatement reaction and the unproductive consumption of the oxidant must be taken into account. Permanganate shows high soil oxidant demand (NOD) due to the reduced species in soil, while CHP and ozone have high decomposition rates in soils. The pH is also a decisive variable for the selection of the oxidant. While permanganate and ozone can be applied in a wide range of pH, CHP, and activated persulfate by metals require acid pH or the addition of chelating agents to obtain a neutral/slightly alkaline media. Soil permeability and changes of this parameter due to the by products formed during the reaction should also be considered in the ISCO application. Finally, total mineralization of the contaminants is not necessary if ISCO is followed by biological treatment or natural attenuation.

---

[1] Chemical Engineering Department. Universidad Complutense de Madrid, Av Complutense s/n, 28040 Madrid, Spain.
[2] Chemical Engineering Department, Universidad de Málaga, Campus de Teatinos s/n, 29071 Málaga, Spain.
\* Corresponding author: aursan@quim.ucm.es

## Introduction

As was commented in the previous chapters, a considerable amount of research has been focused in recent years on trying to remediate contaminated soils instead of destroying or isolating them. Remediation may have several outcomes in terms of soil pollution (Nathanail and Bardos 2004) as is schematized in Fig. 1. Among these, the complete or substantial destruction/degradation of the pollutants is the preferred solution in many cases.

The pollutant abatement can be achieved by both biological and chemical treatments, which are capable of achieving the required level of risk reduction, at reasonable cost and complying with the practical constraints for a particular site. In this sense, biological processes usually require higher times, and in many cases, a combination of processes may offer the most effective remediation.

Chemical treatments can be divided into two categories: chemical reduction and oxidation processes. Chemical oxidation has been widely used for treatment of organic contaminants in soil, groundwaters, and industrial waste waters. In the case of soil remediation, *in situ* treatments are preferred as they require less handling with minimal disruption to activities on site or on adjacent land. However, they are slower and more difficult to put into practice given the difficulty of exposing the decontamination agents to the entire contaminated soil mass. Specifically, *In Situ* Chemical Oxidation (ISCO) is based on the delivery of chemical oxidants to contaminated media in order to destroy the contaminants by converting them to innocuous compounds commonly found in nature (Siegrist et al. 2011). The oxidant chemicals react

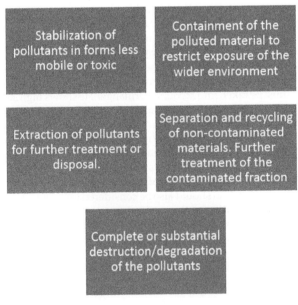

**Fig. 1.** Remediation Outcomes.

with the contaminants, producing innocuous substances such as carbon dioxide, water, and—in the case of chlorinated compounds—inorganic chloride. Contaminants amenable to treatment by ISCO include the following compounds (ITRC 2005):

- benzene, toluene, ethylbenzene, and xylenes (BTEX);
- methyl *tert*-butyl ether (MTBE);
- total petroleum hydrocarbons (TPH);
- chlorinated solvents (ethenes and ethanes);
- polyaromatic hydrocarbons (PAHs);
- polychlorinated biphenyls (PCBs);
- chlorinated benzenes (CBs);
- phenols;
- organic pesticides (insecticides and herbicides); and
- munition constituents (RDX, TNT, HMX, etc.).

In many cases, these compounds are present in the soil as Non Aqueous Phase Liquids (NAPL) (EPA 1991, 1995). Furthermore, the contaminant may be present in several or all phases, such as NAPL, water, entrapped in soil pores, and in the air phase of the vadose zone. Moreover, a plume of dissolved contaminants can appear in the saturated zone, as is represented in Fig. 2.

The main ISCO advantages are the moderate cost for high concentration areas, the short time required, and the effective destruction of the pollutants (Baciocchi 2013; Baciocchi et al. 2014; Tsitonaki et al. 2008). ISCO can also be

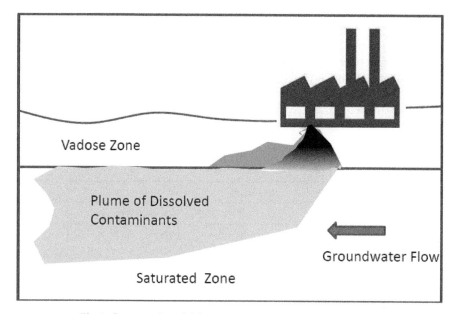

**Fig. 2.** Conceptual model for an organic chemical release into soil.

applied in conjunction with other treatments such as pump-and-treat, soil vapor extraction, surfactants, and bioremediation to break down remaining compounds (Tsitonaki et al. 2008). However, some ISCO drawbacks are the high cost for treatment of dilute plumes, the high consumption of certain oxidants in unproductive reactions with some soils, health, and safety threats due to strong oxidants handling and, occasionally, gas and vapors can also be produced. In addition, ISCO is in general not effective for low permeability zones (EPA 2004; EPA 2006).

In spite of these limitations, ISCO is in many cases less costly and disruptive than other traditional soil treatments such as excavation and incineration, and may be used in applications where the effectiveness of bioremediation is limited by the range of contaminants and/or climate conditions. The oxidants typically used for *in situ* chemical oxidation are:

- Permanganate.
- Ozone.
- Hydrogen peroxide and catalyzed hydrogen peroxide (CHP), this being the most effective. When the catalyst used is an iron salt, the process is referred to as Fenton's reagent.
- Persulfate and activated persulfate. Thermal or metal activation are the most frequently used for *in situ* applications.

The oxidation mechanism can be radicalic or not, depending on the oxidant and the catalyst or activator. Table 1 summarizes the main redox and/or radicalic reactions. Among the four oxidants cited before, permanganate has been used the most widely in ISCO applications at pilot and field scale conditions (Siegrist et al. 2011). Fenton's Reagent usage has been increasing in soil remediation at both the pilot and field scale in the last decade. However, the application of ozone is limited because it needs to be generated *in situ* due to its high instability, and so, it involves significant engineering and safety managements. In contrast, the use of persulfate is emerging.

The chemistry of these oxidants will be described in the following sections, along with a short reflection about the applicability of these oxidants. Although

Table 1. Common oxidants and the most important oxidant reactions.

| Oxidant | Reactions |
|---|---|
| CHP | $H_2O_2 + Fe^{2+} \rightarrow OH^{\cdot} + OH^- + Fe^{3+}$ |
| Ozone | $O_3 + 2H^+ + 2e^- \rightarrow O_2 + 2H_2O$ <br> $O_3 + OH^- \rightarrow O_2 + HO_2^{\cdot}$ |
| Persulfate ($S_2O_8^{2-}$) | $S_2O_8^{2-} + 2e^- \rightarrow 2SO_4^{2-}$ |
| Activated persulfate | $S_2O_8^{2-} + (Fe^2 \text{ or heat}) \rightarrow SO_4^{\cdot-} + (SO_4^{2-} \text{ or } SO_4^{\cdot-})$ <br> $SO_4^{\cdot-} + e^- \rightarrow SO_4^{2-}$ |
| Permanganate | $MnO_4^- + 4H^+ + 3e^- \rightarrow MnO_2(s) + 2H_2O$ |

a brief description of the chemistry of permanganate and ozone will be also offered, the attention will be mainly focused on CHP and activated persulfate, since these oxidants have received more attention in recent years.

## Permanganate as Oxidant

Permanganate is able to oxidize different organic compounds, such as chlorinated solvents (Woo et al. 2010; Mahmoodlu et al. 2014), hydrocarbons (Silva et al. 2011), phenols, and some PAHs in groundwater and soils (Waldemer and Tratnyek 2006; Silva et al. 2009; Liao et al. 2011). In general, it can be established that permanganate is highly reactive with alkene bonds and less reactive with compounds containing single carbon-carbon bonds (saturated hydrocarbons).

Permanganate may be applied to sites as either potassium permanganate (KMnO$_4$) or sodium permanganate (NaMnO$_4$). The advantage of sodium permanganate is that it can be supplied as a solution. In fact, a standard 40% solution is available to the soil remediation market, with low concentrations of heavy metals and impurities. Potassium permanganate is cheaper, but it is yet only available as a solid. Therefore, in order to inject it, it needs to be previously dissolved, which brings a number of adverse side effects such as dust formation. Besides, potassium permanganate is derived from mined potassium ores and can contain different salts and metal impurities (e.g., arsenic, chromium, lead) that could produce some alarm depending on the water quality criteria in the region in which the site is located. In addition, the maximum solubility of potassium permanganate (approximately 6%) is lower than that cited for sodium permanganate (40%). However, potassium permanganate is more widely available and less costly.

All permanganate solutions are dark purple in color and permanganate decomposition can be monitored by evaluating the color of the groundwater. Another remarkable advantage over other oxidants is that permanganate is able to oxidize organics over a wider pH range. In this sense, permanganate reacts primarily through direct electron transfer reactions rather than through radical intermediates. Therefore, the reaction mechanism and number of electrons transferred depends on the pH of the system in such a way that the amount of electrons transferred increases with the pH decrease. The main half reactions can be written as follows (Baciocchi 2013):

$$MnO_4^- + 8H^+ + 5e^- \rightarrow Mn^{2+} + 4\,H_2O \qquad pH < 3.5 \qquad [1]$$

$$MnO_4^- + 2H_2O + 3e^- \rightarrow MnO_2(s) + 4OH^- \qquad 3.5 < pH < 12 \qquad [2]$$

$$MnO_4^- + 1e^- \rightarrow MnO_4^{2-} \qquad pH > 12 \qquad [3]$$

The reaction in Equation [2] is the dominant reaction, at most natural groundwater pH values, except for the cases that pH is significantly altered due to the reaction products, for instance, in the treatment of highly contaminated zones, such as adjacent to NAPL sources, due to the oxidation products of TCE, etc. as it is described in equation [4]:

$$2KMnO_4 + C_2HCl_3 \rightarrow 2CO_2 + 2MnO_2 + 4K^+ + 8H^+ + 12Cl^- \qquad [4]$$

If pH decreases, a mobilization of naturally-occurring metals and metal contaminants that may also be present in the treatment area, could take place. However, this is usually a transient phenomenon, which may be offset by sorption of the metals onto strongly sorbent $MnO_2$ solids that are precipitated as a byproduct of permanganate oxidation. It should also be pointed out that the mineralization of the pollutants does not take place directly, but it appears by an oxidation route. In the case of TCE mentioned before, other organic byproducts have also been detected (Hsu et al. 2006; Liang et al. 2014).

As has been commented, *in situ* applications of permanganate produce solid manganese dioxide that usually appears as a colloid in aqueous systems (Crimi and Siegrist 2004). While manganese oxide is naturally found in soils, this byproduct does not represent an environmental concern, except as a cause of permeability reduction of the soil, mainly nearby the injection wells, which can significantly interfere with oxidant delivery and treatment efficiency (Heiderscheidt et al. 2008).

A good contact between the oxidant and the contaminants is a key point in the success of the treatment. Therefore, a suitable delivery mechanism must be capable of dispersing the oxidant throughout the treatment zone (Baciocchi 2014; Tsitonaki et al. 2008). Multiple injections are often required, with treatment times of several months. In this sense, dissolved permanganate has been delivered to injection or re-circulation wells at concentrations ranging from 100 to 40,000 milligrams per liter (mg/L). Contaminated soils have been successfully oxidized through slurry injection, deep soil mixing, or hydraulic fracturing using concentrated permanganate solutions ranging from 5,000 to 40,000 mg/L or up to 50 percent by weight of solid permanganate (EPA 2004).

On the other hand, permanganate has a high lifetime in the subsoil allowing a more effectively dispersion of the oxidant into the soil and consequently a better contact with adsorbed contaminants. Besides, it normally does not produce heat, steam, and vapors or associated health and safety concerns.

However, the main drawback of permanganate, apart from its selectivity to some contaminant of concern (COC) and the permeability reduction due to the manganese oxide formed, is that permanganate reacts with naturally occurring reduced materials in subsurface porous media (Xu et al. 2009; Cha et al. 2012). This is called natural oxidant demand (NOD). Permanganate NOD is a kinetically controlled process. Usually a permanganate laboratory test is carried out prior to the remediation, in order to quantify how much oxidant the soil consumes. On the other hand, the NOD of soil depends on the permanganate concentration. Therefore, the test must be carried out in the range of concentrations under which the remediation is carried out. In general, it is accepted that when the NOD value exceeds 2 g $MnO_4^-$ per kg of soil, the application of permanganate is no longer cost-effective.

## Ozone as Oxidant

Ozone is an allotrope of oxygen and is more soluble in water than oxygen. It is a strong oxidant and as is shown in Table 1, ozone can react with the COC directly or via hydroxyl radicals, being the radical mechanism much faster than oxidation by ozone itself. It has been widely used in wastewater applications (Loeb 2009).

Direct oxidation involves the oxidation of the COC by ozone as follows (Qui et al. 1999):

$$O_3 + RC = CR \rightarrow RCOCR + O_2 \quad \text{[5]}$$

The radical mechanism is due to the produced hydroxyl radical, $OH^.$, which is a nonselective oxidizer, which attacks organic contaminants and breaks down their carbon-carbon bonds (Da Silva et al. 2006). Ozone may degrade a wide array of organic contaminants. Aromatic compounds, such as BTEX (Bhuyan et al. 2012), PAH's (O'Mahony et al. 2006; Russo et al. 2010), phenols, explosives, PCBs, and pesticides (Zhang et al. 2003; Ikehata et al. 2005), are very reactive towards this oxidant. The mineralization of COC is not direct but oxygenated byproducts are formed, and the number and type of organic intermediates depends on the target pollutant, but in many cases aldehydes, ketones, carboxylic acids are found (Kim et al. 2008; Turhan et al. 2008).

The decomposition of ozone into free radicals is enhanced by an alkaline pH. The exact reaction pathway and radicals taking place in the system, involving ozone and organic compounds, are yet under discussion, but could be grouped as (Parson et al. 2004; ITRC 2005; Siegrist et al. 2011):

$$O_3 + OH^- \rightarrow O_2 + HO_2^. \quad \text{[6]}$$

$$O_3 + HO_2^. \rightarrow OH^. + 2O_2 \quad \text{[7]}$$

$$OH^. + RH \rightarrow R^. + OH^- \quad \text{[8]}$$

$$R^. + O_3 + H_2O \rightarrow ROH + O_2 + OH^. \quad \text{[9]}$$

Since ozone is a gas, it can be used in both saturated and unsaturated parts of the soil. In fact, ozone is probably the most convenient oxidant to employ in the vadose zone. If ozone is used in the unsaturated zone this oxidant is injected as a gas. On the other hand, in the saturated zone, ozone sparging can be performed below the water table or dissolved in water prior being injected. When ozone is sparged in the saturated zone, preferential flow paths can appear due to underground heterogeneities: gas moves upwards and soil usually has a horizontal stratification.

Combinations of hydrogen peroxide and ozone (peroxone) have also been described (Patterson et al. 2013). The general reaction of this system producing the hydroxyl radical can be summarized as follows:

$$2O_3 + H_2O_2 \rightarrow 2OH^{\cdot} + 3O_2 \qquad [10]$$

Since ozone decomposes quickly with a half-life of hours or less, it must be generated on site. To do this, ozone generators producing an air or oxygen stream containing ozone within the 2–10 wt% range are required. This can be a limitation of this oxidant's application. When ozone is introduced as a gas, the application rate is controlled by the equilibrium between gas and liquid phases. When ozone is previously dissolved in water, the aqueous equilibrium ozone concentrations generally range between 5 and 30 mg/L (Langlais et al. 1991).

Ozone delivery and distribution to the targeted zone(s) are critical for effective remediation. Oxidant transport is limited by numerous ozone reactions that occur in the subsurface, as well as geological and hydrogeological properties of the soil such as permeability, heterogeneity, and grain size. Furthermore, since it is quickly decomposed, longer injection times may be required in comparison to those needed for other oxidants. On the other hand, gas generation in the soil is expected when ozone is employed, and this requires the necessary safety consideration in the field applications. Besides, as with other oxidants, carbonate ions, naturally present in the soil, act as radical scavengers, thus increasing the oxidant demand.

Finally, ozone can enhance a further biodegradation of the pollutants due to the presence of oxygen released as a byproduct (Nam et al. 2000).

## CHP as Oxidant

Hydrogen peroxide ($H_2O_2$) is a strong oxidant that can be injected into a contaminated zone to destroy complex organic compounds. This oxidant alone is not able to degrade the COC before decomposition of hydrogen peroxide occurs. However, if a metal salt is added, such as an iron (II) or iron (III) salt, a remarkable increase of the oxidative strength is noticed. As shown in Table 1, this increase is due to the production of hydroxyl radicals ($OH^{\cdot}$). As mentioned in the ozone section, the hydroxyl radical is a powerful and nonselective oxidant, which can react with many organic compounds.

The reaction of iron catalyzed peroxide oxidation was first discovered by H.J.H. Fenton (Fenton 1984) and the iron/peroxide mixture is known as "Fenton's reagent" (FR). This oxidation mixture has been commonly used in the abatement of non-biodegradable organic compounds in wastewater

(Parsons 2004; Neyens et al. 2003; Trapido et al. 2009; Santos et al. 2010) and has also been widely used for the remediation of contaminated soil and groundwater (Watts et al. 2000; Kwan and Voelker 2002; Watts et al. 2006). The use of Fenton's Reagent in the remediation of contaminated soil has proven to be successful in the removal of hydrocarbons (Kang and Hua 2005), PAHs (Jonsson et al. 2007), chlorinated solvents (Teel et al. 2001), ammunition (Li et al. 1997), pesticides (Mecozzi et al. 2006), BTEX waste oil, diesel fuel, etc. (Watts et al. 2000; Pardo et al. 2014a), diesel (Pardo et al. 2014b), and MTEB (Innocenti et al. 2014).

The main reactions taking place in the subsurface by using Fenton's Reagent can be summarized as follows:

$$Fe^{2+} + H_2O_2 \rightarrow Fe^{3+} + OH^- + OH^{\cdot} \qquad [11]$$

$$RH + OH^{\cdot} \rightarrow RHOH^{\cdot} \qquad [12]$$

$$H_2O_2 + OH^{\cdot} \rightarrow HO_2^{\cdot} + H_2O \qquad [13]$$

$$Fe^{3+} + H_2O_2 \rightarrow Fe^{2+} + H^+ + HO_2^{\cdot} \qquad [14]$$

$$Fe^{3+} + HO_2^{\cdot} \rightarrow Fe^{2+} + O_2 + H^+ \qquad [15]$$

$$OH^{\cdot} + Fe^{2+} \rightarrow Fe^{3+} + OH^- \qquad [16]$$

$$HO_2^{\cdot} + Fe^{2+} \rightarrow Fe^{3+} + HO_2^- \qquad [17]$$

Reaction [11] is a fast reaction that generates the hydroxyl radicals. As showed in reactions [12] to [17], chain-propagating and chain-termination sequences take place, involving inorganic and organic radicals. As previously remarked, the COC (symbolized by RH) is not directly mineralized but many oxidized organic intermediates are produced in serial reactions. In general, the more oxygenated the by products, the more biodegradable they are. On the contrary, oxygenated by-products are more recalcitrant against chemical oxidation. Therefore, the oxidized by-products can go to natural attenuation, saving oxidant dosage.

Reaction [11] produces $Fe^{3+}$ and this must be turned to $Fe^{2+}$ to close the catalytic cycle completely. However, the kinetic constant of reaction [14] is two or three orders of magnitude lower than reaction [11] (Neyens et al. 2003), and consequently, a fast depletion of $Fe^{2+}$ is noticed when this species is initially added. Therefore, similar results are usually obtained by adding $Fe^{2+}$ or $Fe^{3+}$ salts (Pardo 2014a).

The optimal pH for the Fenton reaction is about 2.5–3, due to both the higher rate of hydroxyl radicals formation (Emami et al. 2010) and the need of keeping the $Fe^{3+}$ in solution, avoiding the catalytic cycle breaking-off, due to precipitation of the oxidized iron species. Solubility of $Fe^{3+}$ dramatically decreases at neutral-alkaline conditions (Gayer and Woontner 1956), therefore, it is necessary to either lower the pH or use chelating agents to avoid iron precipitation. Common acids, which are used to alter the subsurface pH, are HCl, $H_2SO_4$, and acetic acid. However, these acids can produce undesirable reactions in high-organic soils; the anion used (if chloride) could act as a radical scavenger, and the pH decrease increases the risk of heavy metals mobilization. Another way to increase the iron solubility is to use a chelating agent, in order to avoid the loss of catalyst due to precipitation of iron in the form of $Fe(OH)_3$, at the near-neutral pH found in some soils (Watts et al. 2005; Lewis et al. 2009). When Fenton's Reagent is added together with an organic complexing agent, it is called modified Fenton's Reagent (MFR). While in the USA the most used complexing agent is EDTA, other more biodegradable chelating agents have been recently proposed for this scope (Sun and Pignatello 1992; Lewis et al. 2009; Vicente et al. 2011; Jho et al. 2012; Venny et al. 2012). Citrate complies with these requirements as chelating compound, and its addition, as sodium salt, instead of the acidic form, promotes a near-neutral soil pH during reaction, minimizing the environmental impact of the remediation technique (Vicente et al. 2011). This sodium citrate is also able to keep iron in solution due to its ability of binding ferrous and ferric iron, and can also minimize the nonproductive consumption of hydrogen peroxide (Vicente et al. 2011; Rosas et al. 2014).

The contaminant removal rate of MFR is considerably lower than FR, but the pollutant can be eliminated more efficiently at prolonged times, due to the ability of the chelating agent to keep catalyst in solution for longer periods of time (Rastogi et al. 2009). The use of a neutral pH releases oxidants and oxygen over a longer period, and may promote subsequent aerobic remediation. Besides, chelating agent can solubilize the iron naturally occurring in the soil, avoiding the need of an iron salt addition (Rosas et al. 2014). However, chelating agents can also be attacked by hydroxyl radicals, competing with pollutants (Pardo et al. 2014a,b). Sillanpaa et al. (2011) made a comparison of the oxidation of some chelating agents such as EDTA, NTA, DTPA, and HEDTA by several advanced oxidation processes (AOPs), such as ozonation, Fenton- or Fenton-Like reactions, in aqueous solutions. In this sense, the effectiveness of chelating agents strongly depends on the type of contaminant. In fact, when citrate was used with gasoline type compounds, the total abatement of the COC was achieved before the citrate was oxidized (Pardo et al. 2014a). On the contrary, if diesel is selected as COC, it was noticed than citrate was oxidized at a much higher rate than TPH (Pardo et al. 2014b). These differences can be associated to the fact that diesel contains organic compounds that are more refractory toward chemical oxidation than the organic compounds found in BTEX. Furthermore, the chelating agent oxidation and the associate consumption of the oxidant are much higher in the soil contaminated by diesel, in comparison with the one contaminated with TPH. Consequently, an adequate selection of the chelating agent should take into account the chemistry of both the soil and the COC.

Apart from the pH range, there is one major limitation for the use of Fenton's Reagent, which is the hydrogen peroxide decomposition through unproductive reactions, as is schematized by this equation below:

$$H_2O_2 \xrightarrow{soil} H_2O + \frac{1}{2}O_2 \qquad [18]$$

which do not produce hydroxyl radicals but generate gases that hinder the applicability of this technology (Watts et al. 1999). The soil compounds associated with the decomposition of hydrogen peroxide to water and oxygen are inorganic compounds, such as iron and manganese oxyhydroxides catalysts, as well as other transition metals resulting from mineral dissolution, which are widespread in surface soils (Baciocchi et al. 2003). A strategy applied to minimize this reaction is the use of stabilizers such as phosphate to minimize the formation of gases (Vicente et al. 2011).

In addition, soil organic matter (SOM) and/or radical scavengers (carbonates, chloride, etc.) can act as competitors by reacting with the $HO^\bullet$ radical (Romero et al. 2011):

$$SOM + OH^\bullet \rightarrow SOM_{oxidized} \qquad [19]$$

$$CO_3^{2-} + OH^\bullet \rightarrow CO_3^{\bullet-} + OH^- \qquad [20]$$

The instability of hydrogen peroxide in soil systems can make the delivery of this oxidant difficult for an *in situ* remediation, without decomposing close to the surface. This fact implies that an excess of oxidant with respect to the chemical oxidant is necessary to remove the contaminants from the soil. Besides, injection points should be carefully designed taking into account the lifetime of the oxidant. The occurrence of high SOM, carbonates, or species in the soil decomposing hydrogen peroxide could discourage the use of Fenton's Reagent.

## Activated Persulfate as Oxidant

Persulfate salts dissociate in water to persulfate anions that are strong but relatively stable oxidants as shown in Eq. [21]:

$$S_2O_8^{2-} + 2e^- \rightarrow 2SO_4^{2-} \qquad [21]$$

Persulfate (PS) has drawn increasing attention as an alternative oxidant in the abatement of organic contaminants in the last decade. Persulfate has been recently used for the degradation of organic pollutants in wastewaters (Kusic et al. 2011; Rodriguez et al. 2012; Fang et al. 2013) and is an emerging oxidant for ISCO (Liang and Huang 2008). It has several advantages (Tsitonaki et al. 2010) such as high aqueous solubility, high stability at room temperature, relatively low cost, and benign end products. The most common salt used in ISCO applications is sodium persulfate while potassium persulfate has low solubility and a higher cost.

The use of persulfate as oxidant has kinetic limitations as it reacts much more slowly than other oxidants. However, if the persulfate anion is chemically or thermally activated, it produces the sulfate radical ($SO_4^{\bullet-}$), which is a stronger oxidant ($E° = 2.6$ V). Persulfate can be activated through different routes: thermal, photo activation (Eq. [22]), or activation through reduced metal ions, being $Fe^{2+}$ the most commonly used (Eq. [23]):

$$S_2O_8^{2-} \xrightarrow{\text{heat or } h\nu} 2SO_4^{\cdot-} \qquad [22]$$

$$S_2O_8^{2-} + Fe^{2+} \rightarrow Fe^{3+} + SO_4^{2-} + SO_4^{\cdot-} \qquad [23]$$

Other transition metals that have been found to activate persulfate are $Cu^+$ and $Ag^+$ (Tsitonaki et al. 2010), but these are undesirable activators for remediation applications due to their toxicity and higher cost. In addition, $Mn^{2+}$ was found to be ineffective in producing sulfate radicals (Anipsitakis and Dionysiou 2004). This is an important finding given the abundance of this transition metal in soil and aquifer systems.

The sulfate radical is a very potent oxidizing agent, roughly equivalent to the hydroxyl radical generated using ozone or peroxide, and can attack a wide number of organic pollutants. Reactions involving radicals are quite complex, including free radical chain reactions such as initiation (Eq. [22], [23]), propagation, and termination steps. The main chain propagating reactions are:

$$SO_4^{\cdot-} + RH \rightarrow SO_4^- + R^{\cdot} + H^+ \qquad [24]$$

$$SO_4^{\cdot-} + H_2O \rightarrow SO_4^{2-} + OH^{\cdot} + H^+ \qquad [25]$$

$$SO_4^{\cdot-} + OH^- \rightarrow SO_4^{2-} + OH^{\cdot} \qquad [26]$$

As can be seen, not only sulfate radicals are produced but also hydroxyl radicals appear in the reaction media, and the oxidation of the COC can be due to both sulfate and hydroxyl radicals (Fang et al. 2013).

The termination reaction can also involve undesired non-productive consumption of sulfate radicals due to the scavenging of $SO_4^{\cdot-}$ by $Fe^{2+}$

$$SO_4^{\cdot-} + Fe^{2+} \rightarrow Fe^{3+} + SO_4^{2-} \qquad [27]$$

Because reaction [27] takes place at a high rate adding an excess of $Fe^{2+}$ would produce a loss of the oxidant. Therefore, the amount and delivery strategy of iron addition becomes a key point in the persulfate treatment (Vicente et al. 2011b). In order to solve this disadvantage, the use of zero-valent iron (ZVI) has been proposed in the literature as an alternative to the addition of $Fe^{2+}$. ZVI allows a slow release of $Fe^{2+}$ into the aqueous phase minimizing the extension of reaction [27]. Studies with PS + ZVI concern mainly water treatment (Deng et al. 2014; Liang and Guo 2010; Rodriguez et al. 2014) and to a lower extent soil treatment (Oh and Shin 2014), where a ZVI powder was used as activator of persulfate for the treatment of a diesel-contaminated soil. To solve the suitability of the injection of ZVI in geological porous the use of ZVI in the form of nanoparticles has been proposed (Al-Shamsi and Thomson 2013). Apart from reaction [27], unproductive reactions due to radical scavenging by carbonates and chloride anions, as previously cited for CHP, can also take place.

Persulfate activation by $Fe^{3+}$ has been noticed to occur at much slower rates than activation by $Fe^{2+}$, or is possibly ineffective (Romero et al. 2010). However, Anipsitakis and Dionysiou (2004), reported a mild catalytic activity of $Fe^{3+}$ with persulfate in the degradation of 2,4-dichlorophenol. However, they attributed this to the direct reaction between $Fe^{3+}$ and the aromatic contaminant. Similar findings were obtained by Rodriguez et al. (2014) when using $Fe^{3+}$ with PS in decoloration of an azo dye, where the reduction of $Fe^{3+}$ was attributed to the hydroquinone-quinone-type by-products formed. Therefore, the feasibility of the reduction of $Fe^{3+}$ to $Fe^{2+}$ by COC or oxidation by-products will depend on the chemical nature of both COC and organic by-products, and consequently, these compounds will influence the required amount of $Fe^{2+}$. Ferrous and ferric ions require acidic pH to remain in solution. It may be necessary to lower the pH as with peroxide systems to achieve this environment.

Another approach to activate the sulfate radical is the use of elevated pH. As shown in Eq. [26], a basic solution should increase the production of free hydroxyl radicals. In research, lime has been added to generate an excess of hydroxyl ions and PS is thermally activated from the heat of hydration of the lime to form free sulfate radicals. Recent works have also demonstrated that under alkaline conditions persulfate can decompose a wide range of organic pollutants (Waisner et al. 2008).

A main advantage of using persulfate is that although this oxidant does react with aquifer solids resulting in oxidant consumption, its persistence varies from days to months, depending on the conditions. The impact of persulfate on metal mobility is not well understood. Conceivably, persulfate could impact metal concentrations in groundwater through modification of pH, oxidation of metals, injection of activation amendments, and other mechanisms (Siegrist et al. 2011).

## ISCO Applicability and Comparison Between Oxidants

Critical factors in ISCO applicability are the selection and dosage of the oxidant and the effective distribution of the reagents in the treatment zone. For this, a careful site characterization is required. This implies that both the location and the chemistry of the contamination source, which is usually a Non Aqueous Liquid Phase (NAPL), and the extension and concentration of the plume must be accomplished. If a free or mobile phase (continuous NAPL pools) is present, ISCO is possible, but would require a high oxidant dose. However, ISCO applicability increases if only residual NAPLs (discontinuous or sorbed liquid phases) are present. The presence of mobile free products would require a previous technology (e.g., physical extraction) to reduce the subsurface volume of free mobile residual products. On the other hand, ISCO is a suitable technology for the treatment of plumes with high enough contaminant concentration.

Moreover, a characterization of the hydrogeology and lithography of the site are required because subsurface heterogeneities or preferential flow paths can cause an uneven distribution of the oxidant, resulting in pockets of untreated contaminants. Permeability of the soils is a critical aspect in ISCO application. Intrinsic permeability is a measure of the ability of soil to transmit fluids and can be calculated from hydraulic conductivity measurements. Intrinsic permeability often decreases near injection wells or infiltration galleries and can also decrease as chemical oxidation progresses due to precipitation of inorganic complexes of iron that form during oxidation. In general, the oxidant is more easily distributed in soils with higher soil permeabilities (e.g., sands and gravels) than in soils with lower permeabilities (e.g., clays or silts). ISCO can be successfully applied if the hydraulic conductivity is higher than $3.10^{-7}$ m/s (intrinsic permeability higher than $10^{-13}$ m$^2$) (EPA 2004).

The unproductive consumption of the oxidant by the SOM or other compounds naturally present in the soil must also be evaluated in order to estimate the amount of oxidant required, the location of the injection wells, and the injection sequences. However, the Natural Oxidant Demand (NOD) for a given oxidant depends on the soil components (Liang et al. 2012). Figure 3 shows an average NOD for the oxidants considered in this chapter (Brown 2003).

The decomposition of the oxidant is also a crucial aspect to consider. As can be seen, persulfate and permanganate would be the most stable oxidants in the subsurface (Fig. 4).

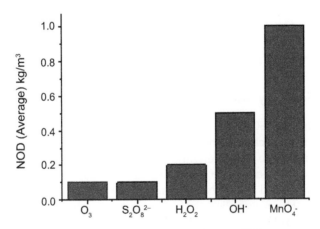

**Fig. 3.** NOD (average) for the different oxidants in ISCO (Brown 2003).

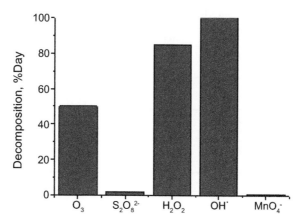

**Fig. 4.** Average Decomposition Rate of different oxidants in ISCO (Brown 2003).

Table 2 reports the oxidant strengths of the main oxidant considered, while Tables 3 and 4 summarize the oxidant effectiveness for hydrocarbon

**Table 2.** Oxidant strenghts (Siegrist et al. 2011).

| Chemical species | Oxidation potential (V) | Relative oxidizing power (Cl2 = 1.0) |
|---|---|---|
| Hydroxyl Radical | 2.8 | 2.0 |
| Sulfate radical | 2.5 | 1.8 |
| Ozone | 2.1 | 1.5 |
| Sodium persulfate | 2.0 | 1.5 |
| Hydrogen peroxide | 1.8 | 1.3 |
| Permanganate (Na/K) | 1.7 | 1.2 |
| Chlorine | 1.4 | 1.0 |
| Oxygen | 1.2 | 0.9 |
| Superoxide ion | 2.4 | 1.8 |

**Table 3.** Oxidant Effectiveness for Hydrocarbon Contaminants (Brown 2003).

| COC | Benzene | TEX | Phenols | PAHs | Munition | PCBs | Pesticides |
|---|---|---|---|---|---|---|---|
| CHP | Good | Good | Good | Medium | Medium | Low | Low |
| KMnO$_4$ | Non Reactive | Good | Good | Good | Good | Low | Medium |
| NaMnO$_4$ | Non Reactive | Good | Good | Good | Good | Low | Medium |
| PS + Fe | Good | Good | Good | Medium | Medium | Low | Medium |
| PS + Heat | Good | Good | Good | Good | Good | Good | Good |
| Ozone | Medium | Medium | Good | Good | Good | Good | Good |

**Table 4.** Oxidant Effectiveness for Chlorinated Volatile Organic Compounds (Brown 2003).

| Oxidant | Amenable | Reluctant | Recalcitrant |
|---|---|---|---|
| CHP | PCE, TCE, DCE, VC, CB | DCA, $CH_2Cl_2$ | TCA, CT, $CHCl_3$ |
| $KMnO_4$ | PCE, CE, DCE, VC | | TCA, CT, $CHCl_3$, DCA, CB, $CH_2Cl_2$ |
| $NaMnO_4$ | PCE, TCE, DCE, VC | | TCA, CT, $CHCl_3$, DCA, CB, $CH_2Cl_2$ |
| PS + Fe | PCE, TCE, DCE, VC, CB | DCA, $CH_2Cl_2$, $CHCl_3$ | TCA, CT |
| PS + Heat | All | | |

and chlorinated contaminants, respectively. The main considerations for *in situ* treatment with ISCO are summarized in Table 5.

Regarding the estimated price for ISCO application, the oxidant and execution costs must be taken into account. A range from about 10 to 100 euros per cubic meter can be found in field applications. In order to reduce the cost with acceptable risk, treatment trains have been proposed (Siegrist et al. 2011) combining ISCO with surfactants (S-ISCO), and/or subsequent natural or stimulated bioremediation.

**Table 5.** Considerations for soil remediation with ISCO.

| | CHP | Ozone | Permanganate | Persulfate |
|---|---|---|---|---|
| Potential Negative Effects | Gas evolution, heat generation, by-products, resolubilization of metals | Gas evolution, by-products, resolubilization of metals | By-products, resolubilization of metals | By-products, resolubilization of metals |
| pH range application | Acid pH Chelating are needed at neutral-alkaline pH (carbonates act as radical scavenger) | Wide pH range, (carbonates act as radical scavenger) | Wide pH range | Wide pH range, Activated by Fe: Chelating are needed (carbonates act as radical scavenger) |
| Persistence | Low Lifetime | Low Lifetime | Stable | Stable |

# References

Al-Shamsi, M.A. and N.R. Thomson. 2013. Treatment of a trichloroethylene source zone using persulfate activated by an emplaced Nano-Pd-Fe-⁰ Zone. Water Air Soil Pollut. 224: 1780.

Anipsitakis, G.P. and D.D. Dionysiou. 2004. Radical generation by the interaction of transition metals with common oxidants. Environ. Sci. Technol. 38: 3705–3712.

Baciocchi, R. 2013. Principles, developments and design criteria of *in situ* chemical oxidation. Water Air Soil Pollut. 224: 1–11.

Baciocchi, R., M.R. Boni and L. D'Aprile. 2003. Hydrogen peroxide lifetime as an indicator of the efficiency of 3-chlorophenol Fenton's and Fenton-like oxidation in soils. J. Hazard. Mater. 96: 305–329.

Baciocchi, R., L. D'Aprile, I. Innocenti, F. Massetti and I. Verginelli. 2014. Development of technical guidelines for the application of in-situ chemical oxidation to groundwater remediation. J. Cleaner Prod. 77: 47–55.

Bhuyan, S.J. and M.R. Latin. 2012. BTEX remediation under challenging site conditions using *In Situ* ozone injection and soil vapor extraction technologies: a case study. Soil Sediment Contam. 21: 545–556.

Brown, R.A. 2003. *In Situ* Chemical Oxidation: Performance, Practice, and Pitfalls AFCEE Technology Transfer Workshop, San Antonio, Texas, February.

Cha, K.Y., M. Crimi, M.A. Urynowicz and R.C. Borden. 2012. Kinetics of permanganate consumption by natural oxidant demand in aquifer solids. Environ. Eng. Sci. 29: 646–653.

Crimi, M.L. and R.L. Siegrist. 2004. Impact of reaction conditions on $MnO_2$ genesis during permanganate oxidation. J. Environ. Eng. 130: 562–572.

Da Silva, L.M. and W.F. Jardim. 2006. Trends and strategies of ozone application in environmental problems. Quimica Nova 29: 310–317.

Deng, J., Y. Shao, N. Gao, Y. Deng, C. Tan and S. Zhou. 2014. Zero-valent iron/persulfate(Fe-0/PS) oxidation acetaminophen in water. Inter. J. Environ. Sci. Technol. 11: 881–890.

Emami, F., A.R. Tehrani-Bagha, K. Gharanjig and F.M. Menger. 2010. Kinetic study of the factors controlling Fenton-promoted destruction of a non-biodegradable dye. Desalination 257: 124–128.

[EPA] United States Environmental Protection Agency 1991. A Ground Water Issue: Dense Nonaqueous Phase Liquids, EPA 540/4-91/002, Washington, DC, March.

[EPA] United States Environmental Protection Agency 1995. A Ground Water Issue: Light Nonaqueous Phase Liquids, EPA 540/5-95/500, Washington, DC, July.

[EPA] United States Environmental Protection Agency 2004. How To Evaluate Alternative Cleanup Technologies For Underground Storage Tank Sites: A Guide For Corrective Action Plan Reviewers, EPA 510-R-04-002, Washington, DC, May.

[EPA] United States Environmental Protection Agency 2006. *In-Situ* Chemical Oxidation–Engineering Issue, EPA/600/R-06/072, Washington, DC, August.

Fang, G.D., D.D. Dionysiou, D.M. Zhou, Y. Wang, X.D. Zhu, J.X. Fan, L. Cang and Y.J. Wang. 2013. Transformation of polychlorinated biphenyls by persulfate at ambient temperature. Chemosphere 90: 1573–1580.

Fenton, H.J.H. 1984. Oxidation of tartaric acid in presence of iron. J. Chem. Soc. Trans. 65: 899–910.

Gaer, K.H. and L. Woontner. 1956. The solubility of ferrous hydroxide and ferric hydroxide in acidic and basic media at 25°. J. Phys. Chem. 60: 1569–1571.

Heiderscheidt, J.L., R.L. Siegrist and T.H. Illangasekare. 2008. Intermediate-scale 2-D experimental investigation of *in situ* chemical oxidation using potassium permanganate for remediation of complex DNAPL source zones. J. Contam. Hydrol. 102: 3–16.

Hsu, C.S., M.C. Lu, J. Anotai and Y.Y. Huang. 2006. Chemical oxidation of TCE-contaminated groundwater by permanganate. Fresen. Environ. Bull. 15: 125–128.

Ikehata, K. and M.G. El-Din. 2005. Aqueous pesticide degradation by ozonation and ozone-based advanced oxidation processes: A review (Part II). Ozone-Sci. & Eng. 27: 173–202.

Innocenti, I., I. Verginelli, F. Massetti, D. Piscitelli, R. Gavasci and R. Baciocchi. 2014. Pilot-scale ISCO treatment of a MtBE contaminated site using a Fenton-like process. Sci. Total Environ. 485: 726–738.

[ITRC] Interstate Technology & Regulatory Council 2005. Technical and Regulatory Guidance for *In Situ* Chemical Oxidation of Contaminated Soil and Groundwater. 2nd Edition. Available at: www.itrcweb.org (accessed 07.20.2014).

Jho, E., N. Singhal and S. Turner. 2012. Tetrachloroethylene and hexachloroethane degradation in Fe(III) and Fe(III)–citrate catalyzed Fenton systems. J. Chem. Technol. Biotechnol. 87: 1179–1186.

Jonsson, S., Y. Persson, S. Frankki, B. Van Bavel, S. Lundstedt, P. Haglund and M. Tysklind. 2007. Degradation of polycyclic aromatic hydrocarbons (PAHs) in contaminated soils by

Fenton's reagent: a multivariate evaluation of the importance of soil characteristics and PAH properties. J. Hazard. Mater. 149: 86–96.

Kang, N. and I. Hua. 2005. Enhanced chemical oxidation of aromatic hydrocarbons in soil systems. Chemosphere 61: 909–922

Kim, I.H., H. Tanaka, T. Iwasaki, T. Takubo, T. Morioka and Y. Kato. 2008. Classification of the degradability of 30 pharmaceuticals in water with ozone, UV and $H_2O_2$. Water Sci. Technol. 57: 195–200.

Kusic, H., I. Peternel, U. Sime, N. Koprivanac, T. Bolanca, S. Papic and A.L. Bozic. 2011. Modeling of iron activated persulfate oxidation treating reactive azo dye in water matrix. Chem. Eng. J. 172: 109–121.

Kwan, W.P. and B.M. Voelker. 2002. Decomposition of hydrogen peroxide and organic compounds in the presence of dissolved iron and ferrihydrite. Environ. Sci. Technol. 36: 1467–1476.

Langlais, B., D. Reckhow and D. Brink. 1991. Ozone in Water Treatment. Chelsea, Mich.

Lewis, S., A. Lynch, L. Bachas, S. Hampson, L. Ormsbee and D. Bhattacharyya. 2009. Chelate-modified Fenton reaction for the degradation of trichloroethylene in aqueous and two-phase systems. Environ. Eng. Sci. 26: 849–859.

Li, Z.M., S.D. Comfort and P.J. Shea. 1997. Destruction of 2,4,6-trinitrotoluene (TNT) by Fenton oxidation. J. Environ. Qual. 26: 480–487.

Liang, C. and C. Huang. 2008. Potential for activated persulfate degradation of BTEX contamination. Water Res. 42: 4091–4100.

Liang, C.J. and Y. Guo. 2010. Mass transfer and chemical oxidation of naphthalene particles with zerovalent iron activated persulfate. Environ. Sci. Technol. 44: 8203–8208.

Liang, C.J., Y.C. Chien and Y.L. Lin. 2012. Impacts of ISCO persulfate, peroxide and permanganate oxidants on soils: soil oxidant demand and soil properties. Soil Sediment Contam. 21: 701–719.

Liang, S.H., K.F. Chen, C.S. Wu, Y.H. Lin and C.M. Kao. 2014. Development of KMnO4-releasing composites for *in situ* chemical oxidation of TCE-contaminated groundwater. Water Res. 54: 149–158.

Liao, X.Y., D. Zhao and X.L. Yan. 2011. Determination of potassium permanganate demand variation with depth for oxidation-remediation of soils from a PAHs-contaminated coking plant. J. Hazard. Mater. 193: 164–170.

Loeb, B.L. 2009. Ozone: science & engineering thirty years of progress. Ozone-Sci. & Eng. 31: 379–392.

Mahmoodlu, M.G., S.M. Hassanizadeh and N. Hartog. 2014. Evaluation of the kinetic oxidation of aqueous volatile organic compounds by permanganate. Sci. Total Environ. 485-486: 755–763.

Mecozzi, R., L. Di Palma and C. Merli. 2006. Experimental *in situ* chemical peroxidation of atrazine in contaminated soil. Chemosphere 62: 1481–1489.

Nam, K. and J.J. Kukor. 2000. Combined ozonation and biodegradation for remediation of mixtures of polycyclic aromatic hydrocarbons in soil. Biodegradation 11: 1–9.

Nathanail, C.P. and R.P. Bardos. 2004. Geology for Contaminated Land, in Reclamation of Contaminated Land, John Wiley & Sons, Ltd, Chichester, UK. doi: 10.1002/0470020954.ch4.

Neyens, E. and J. Baeyens. 2003. A review of classic Fenton's peroxidation as an advanced oxidation technique. J. Hazard. Mater. 98: 33–50.

Oh, S.Y. and D.S. Shin. 2014. Treatment of diesel-contaminated soil by fenton and persulfate oxidation with zero-valent iron. Soil Sediment Contam. 23: 180–193.

O'Mahony, M.M., A.D.W. Dobson, J.D. Barnes and I. Singleton. 2006. The use of ozone in the remediation of polycyclic aromatic hydrocarbon contaminated soil. Chemosphere 63: 307–314.

Pardo, F., J.M. Rosas, A. Santos and A. Romero. 2014a. Remediation of soil contaminated by NAPLs using modified Fenton reagent: application to gasoline type compounds. J. Chem. Tech. Biotech. doi: 10.1002/jctb.4373.

Pardo, F., J.M. Rosas, A. Santos and A. Romero. 2014b. Remediation of a biodiesel blend-contaminated soil with activated persulfate by different sources of iron. Water, Air and Soil Pollution, doi 10.1007/s11356-014-2997-2.

Parsons, S. 2004. Advanced Oxidation Processes for Water and Wastewater Treatment. IWA Publishing, London.

Patterson, C.L., F. Cadena, R. Sinha, D.K. Ngo-Kidd, A. Ghassemi and E.R. Krishnan. 2013. Field treatment of MTBE-Contaminated Groundwater Using Ozone/UV Oxidation. Ground Water Monit. R. 33: 44–52.

Qui, Y., M. Zappi, C. Kuo and E. Fleming. 1999. A kinetic and mechanistic study of the ozonation of dichlorophenols in aqueous solutions. J. Environ. Eng. 125: 441–450.

Rastogi, A., S.R. Al-Abed and D.D. Dionysiou. 2009. Effect of inorganic, synthetic and naturally occurring chelating agents on Fe(II) mediated advanced oxidation of chlorophenols. Water Res. 43: 684–694.

Rinaldi, A. and M.R. Da Silva. 2011. Degradation of BTX in contaminated soil by using Hydrogen Peroxide ($H_2O_2$) and Potassium Permanganate ($KMnO_4$). Water Air Soil Pollut. 217: 245–254.

Rodriguez, S., A. Santos, A. Romero and F. Vicente. 2012. Kinetic of oxidation and mineralization of priority and emerging pollutants by activated persulfate. Chem. Eng. J. 172: 225–234.

Rodriguez, S., L. Vasquez, D. Costa, A. Romero and A. Santos. 2014. Oxidation of Orange G by persulfate activated by Fe(II), Fe(III) and zero valent iron (ZVI). Chemosphere 101: 86–92.

Romero, A., A. Santos, F. Vicente and C. Gonzalez. 2010. Diuron abatement using activated persulphate: Effect of pH, Fe(II) and oxidant dosage. Chem. Eng. J. 162: 257–265.

Romero, A., A. Santos, T. Cordero, J. Rodriguez-Mirasol, J.M. Rosas and F. Vicente. 2011. Soil remediation by Fenton-like process: phenol removal and soil organic matter modification. Chem. Eng. J. 170: 36–43.

Rosas, J.M., F. Vicente, E.G. Saguillo, A. Santos and A. Romero. 2014. Remediation of soil polluted with herbicides by Fenton-like reaction: Kinetic model of diuron degradation. Appl. Catal. B: Environ. 144: 252–260.

Russo, L., L. Rizzo and V. Belgiorno. 2010. PAHs contaminated soils remediation by ozone oxidation. Desalin. Water Treat. 23: 161–172.

Santos, A., P. Yustos, S. Rodriguez and A. Romero. 2010. Mineralization lumping kinetic model for abatement of organic pollutants using Fenton's reagent. Catal. Today 151: 89–93.

Siegrist, L., M. Crimi and T.K. Simpkin. 2011. *In Situ* Chemical Oxidation for Groundwater Remediation. Springer Science+Business Media, LLC, New York, New York. A reference book in the SERDP/ESTCP Remediation Technology Monograph Series, C.H. Ward (Series ed). ~ 700p.

Sillanpaa, M.E.T., T.A. Kurniawan and W.H. Lo. 2011. Degradation of chelating agents in aqueous solution using advanced oxidation process (AOP). Chemosphere 83: 1443–1460.

Silva, P.T.D.E., V.L. Da Silva, B.D. Neto and M.O. Simmonnot. 2009. Potassium permanganate oxidation of phenanthrene and pyrene in contaminated soils. J. Hazard. Mater. 168: 1269–1273.

Sun, Y. and J.J. Pignatello. 1992. Chemical treatment of pesticide wastes evaluation of Fe(II1) chelates for catalytic hydrogen Peroxide Oxidation of 2,4-D at Circumneutral pH. J. Agric. Food Chem. 40: 322–327.

Teel, A.L. C.R. Warberg, D.A. Atkinson and R.J. Watts. 2001. Comparison of mineral and soluble iron Fenton's catalysts for the treatment of trichloroethylene. Water Res. 35: 977–984.

Trapido, M., N. Kulik, A. Goi, Y. Veressinina and R. Munter. 2009. Fenton treatment efficacy for the purification of different kinds of wastewater. Water Sci. Technol. 60: 1795–1801.

Tsitonaki, A. and P.L. Bjerg. 2008. *In Situ* Chemical Oxidation—Afværgeteknologier—State of the Art. Schæffergården, Gentofte 22. October.

Tsitonaki, A., B. Petri, M. Crimi, H. Mosbaek, R.L. Siegristand and P.L. Bjerg. 2010. *In situ* chemical oxidation of contaminated soil and groundwater using persulfate: a review. Critical Reviews in Environ. Sci. and Technol. 40: 55–91.

Turhan, K. and S. Uzman. 2008. Removal of phenol from water using ozone. Desalination 229: 257–263.

Venny Gan, S. and H.K. Ng. 2012. Inorganic chelated modified-Fenton treatment of polycyclic aromatic hydrocarbon (PAH)-contaminated soils. Chem. Eng. J. 180: 1–8.

Vicente, F., J.M. Rosas, A. Santos and A. Romero. 2011a. Improvement soil remediation by using stabilizers and chelating agents in a Fenton-like process. Chem. Eng. J. 172: 689–697.

Vicente, F., A. Santos, A. Romero and S. Rodriguez. 2011b. Kinetic study of diuron oxidation and mineralization by persulphate: effects of temperature, oxidant concentration and iron dosage method. Chem. Eng. J. 170: 127–135.

Waisner, S., V.F. Medina, A.B. Morrow and C.C. Nestler. 2008. Evaluation of chemical treatments for a mixed contaminant soil. J. Environ. Eng. 134: 743–749.

Waldemer, R.H. and P.G. Tratnyek. 2006. Kinetics of contaminant degradation by permanganate. Environ. Sci. Technol. 40: 1055–1061.

Watts, R.J. and A.L. Teel. 2005. Chemistry of modified Fenton's reagent (catalyzed $H_2O_2$ propagations-CHP) for *in situ* soil and groundwater remediation. J. Environ. Eng. 131: 612–622.

Watts, R.J. and A.L. Teel. 2006. Treatment of contaminated soils and groundwater using ISCO. Pract. Per. Haz. Toxic Rad. Waste Mgmt. 10: 2–9.

Watts, R.J., M.K. Foget, S.H. Kong and A.L. Teel. 1999. Hydrogen peroxide decomposition in model subsurface systems. J. Hazard. Mater. 69: 229–243.

Watts, R.J., D.R. Haller, A.P. Jones and A.L. Teel. 2000. A foundation for the risk-based treatment of gasoline-contaminated soils using modified Fenton's reactions. J. Hazard. Mater. 76: 73–89.

Woo, N.C., S.G. Hyun, W.W. Park, E.S. Lee and F.W. Schwartz. 2010. Characteristics of permanganate oxidation of TCE at low reagent concentrations. Environ. Technol. 30: 1337–1342.

Xu, X.Y. and N.R. Thomson. 2009. A long-term bench-scale investigation of permanganate consumption by aquifer materials. J. Contaminant Hydrol. 110: 73–86.

Zhang, H., M.H. Sung and C.P. Huang. 2003. Mathematical model of *in situ* ozonation for the remediation of 2-chlorophenol contaminated soil. Chinese J. Chem. Eng. 11: 555–558.

# CHAPTER 6

# Soil Contamination and Life Cycle Assessment

*Florinda Figueiredo Martins*

## ABSTRACT

Nowadays there are many competing uses for soil: forests, agriculture, urbanization, waste management operations such as landfill, industry and others that make soil a much requested resource. Due to its essential role in human activities and survival, soil is a resource worthwhile to be protected and preserved for present and future generations. However there is a huge number of contaminated sites worldwide. Erosion, loss of organic matter, salinisation, contamination, sealing are among the most common problems found in soil, which have negative impacts on ecosystems, climate, resources, human health and in economies. In order to solve the problem of soil contamination many techniques have been developed. The assessment of environment impacts is a very important aspect in what concerns soil management and Life Cycle Assessment (LCA) can be a very useful tool. In fact LCA have been used for studying a wide range of remediation technologies such as land filling, stabilization/solidification, etc. Other important aspects are that existing sustainability assessment in this area uses LCA methodology to compare remediation technologies and that LCA can help in the decision-making process. This chapter explores all these issues including the LCA application to soil contamination and its constraints.

REQUIMTE/LAQV, Instituto Superior de Engenharia do Porto, Instituto Politécnico do Porto, Rua Dr. António Bernardino de Almeida 431, 4200-072 Porto, Portugal.
Email: ffm@isep.ipp.pt

## Introduction

### *Soil*

Soil is the top layer of earth's crust and is constituted by minerals, organic matter, water, air, and living organisms. It constitutes an extreme complex matrix and a fragile resource and its formation is a slow process and for that reason it can be considered a non-renewable resource. Soil is essential to many human activities: agriculture, raw materials provider, ecosystems survival and biodiversity maintenance, cycle nutrients, carbon sink, etc. Other aspects concerning human well-being, as access to safe water and food, free of contaminants, and cultural and spiritual components connected to landscapes and ecosystems are also very important. Nowadays there are many competing uses for soil: forests, agriculture, urbanization, waste management operations such as landfill, industry, transport, and others that make soil a much requested resource. Due to its functions and its role in human activities and survival, soil is a resource worthwhile to be protected and preserved for present and future generations. Despite its recognized importance, soil is degrading at a global level and its degradation is a problem that many regions and countries are now addressing, including the European Union. Erosion, loss of organic matter, salinization, contamination, and sealing are among the most common problems found in soil management. All these issues have negative impacts on ecosystems, climate, resources, human health, and economies. These problems demand a more sustainable way of managing this resource taking in consideration that local problems generally tend to increase pressure elsewhere.

In most Member States of the European Union there is no specific legislation concerning soil protection, although other European policies and legislation, concerning agriculture, water, waste, industrial pollution, etc., contributed indirectly to soil protection. In 2006, the Commission adopted the Soil Thematic Strategy intending to protect soils. The Soil Thematic Strategy has four pillars, namely awareness raising, research, integration, and legislation (European Commission 2006). In this context the proposal for a Directive was a key component of this strategy since it would provide measures to identify problems, to prevent soil degradation and to remediate polluted or degraded soil. However in May 2014, the proposal for a Soil Directive was withdrawn while the Seventh Environment Action Programme, that entered into force on 17 January 2014, states that soil degradation is a serious challenge. The priority objective "To protect, conserve and enhance the Union's natural capital" of the Seventh Environment Action Programme, 'Living well, within the limits of our planet', it is the recognition that unsustainable use of land in EU is affecting ecosystems services, biodiversity, increasing vulnerability to climate change and natural disasters, enhancing soil degradation, and desertification. Major problems are related to soil erosion by water, soil contamination, and soil sealing. The number of sites that are thought to be contaminated is huge. Identification and assessment of these sites is essential to prevent environmental, economic, social, and health risks. Land is continuously taken for housing, industry, transport, etc., but the objective set for 2050 is 'no net land take' (European Parliament and Council 2013).

Soil management and protection is at the present also a concern for several countries outside the European Union. Australia has recently launched a policy discussion paper to help improve the sustainable management of Australian soil resources. It reinforces the idea of the lack of recognition and importance of the linkages between soil management and other important environmental problems such as climate change, water management, food security, biodiversity, etc. (Campbell 2008).

Due to the rapid economic development of China, soil pollution has been getting worse and in some areas it is a serious problem. Soil degradation, one of the main problems in China, includes erosion, desertification, salting, sterility, and pollution. They are now in what they call the third stage where priority is given to soil pollution prevention and control (CCICED 2010).

In the United States of America, a comprehensive soil protection strategy does not exist. A number of Federal and State authorities manage the programs and policies about soils. It involves contaminated land, agriculture, forest management, or watershed protection.

These concerns about soil degradation are global since large areas have been degraded or become deserts and the estimates point to 20% of the earth's land being degraded. Critical locations are Africa, South of the equator, South-East Asia, and southern China. Global evolution trends of population, energy demand, food and water demand are expected to increase the need for land. To meet future demands 175 to 220 million ha of additional cropland will be necessary (UNCCD 2012). A holistic approach that considers all the elements and the connections between land, energy, food, and water should be used.

At the United Nations Conference on Sustainable Development (Rio + 20) a new sustainable development goal was discussed to secure the continuing availability of productive land for present and future generations, achieving a land-degradation-neutral world, namely the Zero Net Land Degradation. The objective is to maintain or improve the condition of our land resources and can be achieved by rehabilitation through forest and landscape restoration.

### Contaminated Sites

In 2011–2012 a study was organized by the JRC (Joint Research Centre) European Soil Data Centre and the data request was sent to the 32 EEA (European Environmental Agency) member countries and the seven EEA cooperating countries, which include the 28 European Union Member States.

The indicators used were related to core issues such as the estimative of the extension of soil contamination, progress achieved in the management and control of local soil contamination, identification of sectors that contribute the most to soil contamination, identification of main contaminants, and how much of the public budget is used.

According to this study local contamination was estimated at 2.5 million potentially contaminated sites[1] in Europe of which about 14% (340,000 sites) are expected to be contaminated and will require remediation. One third of these 340,000 contaminated sites were already identified and about 15% have been remediated (Liedekerke et al. 2014).

This study considers four steps in the management of local soil contamination, namely site identification, preliminary investigations, main site investigations, and implementation of risk reduction measures. The results showed that one third of the countries had developed extensive work in mapping polluting activities and potentially contaminated sites and that seven countries were finalizing this step. The majority of countries (28 out of the 39 countries) maintain comprehensive inventories for contaminated sites. In 17 countries, there are policy targets for the management of contaminated sites, however there is no European target concerning this matter (Liedekerke et al. 2014).

The soil contaminants referred were Chlorinated Hydrocarbons (CHC), Mineral oil, Polycyclic Aromatic Hydrocarbons (PAH), Heavy metals, Phenols, Cyanide, Aromatic Hydrocarbons (BTEX), Polychlorinated Biphenyls (PCB), and pesticides. The most frequent contaminants are mineral oil and heavy metals (Table 1) (Panagos et al. 2013).

**Table 1.** Contaminants affecting soil.

| Contaminant | Percentage |
|---|---|
| CHC | 8.3 |
| Mineral oil | 23.8 |
| PAH | 10.9 |
| Heavy metals | 34.8 |
| Phenols | 1.3 |
| Cyanides | 1.1 |
| BTEX | 10.2 |
| Others | 9.3 |

Activities referred as relevant sources of local contamination are waste disposal and treatment, industrial and commercial activities, storage, transport spills on land, military, and nuclear operations (Table 2). The production sectors contribute the most to local soil contamination; on the other hand, for the service sector, gasoline stations are the most reported source of soil contamination (Liedekerke et al. 2014).

---

[1] Potentially Contaminated Site refers to sites where unacceptable soil contamination is suspected but not verified.

**Table-2.** Sources of local soil contamination.[2]

| Key sources | Average[3] % | Sources relevant for soil and groundwater contamination |
|---|---|---|
| Waste disposal and treatment | 38.1 | Municipal waste disposal |
| | | Industrial waste disposal |
| Industrial and commercial activities | 34.0 | Mining |
| | | Oil extraction and production |
| | | Power plants |
| Military | 3.4 | Military sites |
| | | War affected zones |
| Storage | 10.7 | Oil storage |
| | | Obsolete chemicals storage |
| | | Other storages |
| Transport spills on land | 7.9 | Oil spills sites |
| | | Other hazardous substance spills sites |
| Nuclear | 0.1 | Nuclear operations |
| Other | 8.1 | |

[2] Data from (Liedekerke et al. 2014).
[3] Based on 22 countries/regions.

According to the same study the most used remediation technique has been the excavation of contaminated soil and its disposal in landfills.

In the United States of America, the Department of Energy (DOE) manages the groundwater and soil remediation effort. The amounts involved are quite large since inventory at the DOE sites includes 6.5 trillion liters of contaminated groundwater and 40 million cubic meters of soil and debris contaminated with radionuclides, metals, and organics. New technologies and remediation approaches are being tested to reduce risk and life-cycle cleanup costs (US Department of Energy 2014).

In the USA there is The National Priorities List (NPL) which is a list of US's priorities among the known or threatened releases of hazardous substances, pollutants, or contaminants throughout the mentioned country. Its main purpose is to help the EPA to determine which sites need to be further investigated, to assess the risks related to public health and environment. The NPL is divided in two sections, sites generally evaluated and cleaned up by EPA through the General Superfund Section and sites owned or operated by other federal agencies. Nevertheless, EPA is responsible for preparing a

Hazard Ranking System. This list is updated periodically. By June 18, 2014 the proposed sites were 51, the final sites 1321 and deleted sites[4] 380 (EPA 2014).

Information about the BRIC countries (Brazil, Russia, India, and China) is relatively scarce. Recent news about China, revealed by a report previously classified as a state secret, draw a very concerning scenario, revealing that about a fifth of China's soil is contaminated due to the enormous industrial development. Some regions have suffered deteriorating land quality and serious soil pollution.

Pollution sources in China include industrial, domestic waste from urban residents, agricultural chemicals, and waste from breeding of livestock and poultry. There are old and new pollutants and inorganic-organic chemical combinations (CCICED 2010).

In Latin America one of the regions with higher levels of urbanization a project, INTEGRATION—Integrated Urban Development, was developed during four years (2009–2012). This project was coordinated by the city of Stuttgart (Germany) and involved the cities of São Paulo and Rio de Janeiro (Brazil), Bogotá (Colombia), Quito (Equator), Guadalajara, and the Chihuahua state (México) and its scope was the integrated development of degraded and potentially contaminated urban areas (Marker 2013).

## Methods Used for Soil Remediation

For solving the problem of contaminated soil there are three approaches that can be applied, in the first one the soil is excavated and treated or disposed, in the second soil is left in the ground and treated there, and in the third the soil is left in the ground and contained to prevent further contamination and risks to environment and humans. Very frequently the treatments are divided in *in situ* when the soil is left in the ground and in *ex situ* when the soil is excavated and then treated.

One of the most common solutions is containment. In this type of solution no pollutant is removed or treated, but instead the contaminated area is isolated to prevent migration of contaminants and like this the possibility of exposure and the risk are reduced. Several technologies are applied such as caps which are top barriers and are often used to minimize the generation of leachate by preventing surface water infiltration. They can also be used together with other technologies such as barrier walls, groundwater pump-and-treat, and *in situ* treatment.

Barrier walls can be of several types, namely soil-bentonite, soil-cement-bentonite, cement-bentonite, sheet pile (steel or high-density polyethylene), and clay barriers. Before deciding the material of the wall some tests should be performed to evaluate the material's chemical stability in relation to the site's existing compounds and conditions (EPA 1992, 1998).

---

[4] May occur once all response actions are complete and all cleanup goals have been achieved.

The existence of permeable reactive barriers is also important to mention. Inside this kind of barriers, there are reactive materials (usually zero-valent iron) through which the contaminants plume flows under natural gradient converting contaminants to non-toxic compounds or immobilizing them.

Improvements in this area allowed reaching depths much higher than previously through various soils. The choice of a method for a particular site depends on the site characteristics, the permeability and depth required, and of course, costs involved and nearby surroundings (EPA 1998).

Another possibility is to apply immobilization technologies, namely physicochemical methods such as solidification and stabilization and thermal methods such as vitrification.

Solidification is related to an encapsulation process to obtain a solid matrix, reducing and restricting contaminant migration and leaching processes by decreasing the area of exposure. It is usually accomplished by a chemical reaction between waste and solidifying agents such as, for example, cement. The stabilization process also involves chemical reactions but they are used to reduce, for example, the solubility of hazardous materials, thus reducing its leachability (EPA 2013).

Solidification and stabilization processes can be implemented either *in situ* or *ex situ* and when they involve the use of very high temperatures, the wastes can be vitrified, immobilizing contaminants.

Treatments can be divided into three major categories: physical/chemical, biological, and thermal (Liedekerke et al. 2014).

In physical/chemical treatments, several options can be identified such as soil flushing, soil washing, chemical extraction, soil vapor extraction, chemical oxidation/reduction, and chemical dehalogenation.

Soil flushing is an *in situ* process where the contaminated area is flooded through the injection or infiltration of water, a solution (acidic solutions, basic solutions, surfactants etc.) or gaseous mixtures to promote the mobilization of the contaminant by solubilization, formation of emulsions, or even chemical reactions. The fluid containing the contaminants can be removed from a groundwater system for disposal, recirculation, or on-site treatment and reinjection or can be left in place depending on the characteristics of contaminants and fluid used (EPA 1993).

Soil washing on the other hand is an *ex situ* process that uses chemical and physical extraction and separation processes to remove several contaminants. It basically consists of four steps, namely excavation, pretreatment of soil such as screening, washing the soil with a fluid, which is generally water-based and can contain chemical additives, and recovery of clean soil (EPA 1993).

In chemical extraction, contaminants are separated from the soil and physical separation is also used in the process. Two main extraction processes are acid extraction that uses hydrochloric acid (extraction of heavy metals) and solvent extraction that uses organic solvents to remove organic compounds. In the first case, soils are afterwards rinsed with water, and the extraction solution and rinse water are treated. The heavy metals can potentially be recovered.

When solvents are used there are generally other technologies associated to finish the treatment such as soil washing, solidification/stabilization, etc.

In soil vapor extraction, vacuum is applied through a system of wells or vents placed in contaminated areas and like this a flow of gas is established and the contaminants' vapors are extracted and treated *ex situ*. It is usually used to extract volatile compounds (EPA 1994).

In electrokinetic processes, two electrodes are placed in the ground on each side of the contaminated area and then a low voltage direct current is applied causing the contaminants' migration towards the electrodes. The contaminants can afterwards be extracted to a recovery system. This process can include the use of surfactants and other chemical agents to enhance removal of contaminants (EPA 1995).

Chemical oxidation consists of redox reactions which involve the transfer of electrons from one chemical species to another, converting contaminants into less harmful chemical species. Common oxidants are potassium or sodium permanganate, Fenton's catalyzed hydrogen peroxide, hydrogen peroxide, ozone, and sodium persulfate. Chemical reduction also uses redox reactions and Zero-Valent Iron (ZVI) is usually used. It is an *in situ* process and involves the injection of quantities of iron powder directly into contaminated areas.

Biological treatment generally uses microorganisms to degrade the contaminants under aerobic or anaerobic conditions. Bioremediation can be achieved using indigenous microorganisms present in the soil or by a more elaborated process through the addition of air, organic substrates, nutrients, or even microbial cultures to reach the desired degradation. It is also possible to use phytoremediation. In phytoremediation, generally plants are used as collectors of contaminants for the remediation of contaminated sites with organic or inorganic pollutants.

The most common thermal treatment processes are thermal desorption and incineration. *Ex situ* thermal desorption has two phases, in the first one heat is applied to the contaminated material to vaporize contaminants leading to a gas stream which is then treated. Incineration is a process for burning materials which uses high temperatures in order to destroy contaminants. Vitrification can also be considered a thermal process and it uses electrical power to heat and melt soil with contaminants. When the molten material cools it forms a hard, chemically inert, glass-like product with low leaching characteristics (EPA 1994).

Nowadays nanoparticles are also used, such as zero-valent iron nanoparticles (nZVI) (Carroll et al. 2013). The nZVI has several advantages over the micro-scale particles, namely the increase in the degradation reaction rate, decrease of the reductant dosage, etc.

In natural attenuation, natural processes are responsible for the improvements in soil and groundwater. In order to obtain efficient cleanup the right condition must exist underground and the source of pollution must be removed.

The EPA in the United States worked with ASTM International to develop a standard to encourage greener practices for contaminated sites cleanup. The Clean and Green Policy for contaminated sites has as core goals to minimize total energy use and maximize use of renewable energy, minimize air emissions and greenhouse gas generation, minimize water use and impacts to water resources, reduce, reuse, and recycle materials and wastes, and support environmentally-sustainable reuse of remediated land (EPA 2010). When selecting a method or processes for site cleanup these goals should be considered.

## Life Cycle Assessment (LCA)

### *LCA Framework*

LCA is a methodology used to assess several environmental aspects of a product or service throughout its life. LCA is performed in four major phases, namely goal and scope definition, inventory analysis, impact assessment, and interpretation of results (Varanda et al. 2011) based on international standards (ISO 14040).

The first step is to define the objective of the LCA study as well as the functional unit and the reference flows. Then it is necessary to define the product system, set the system boundaries (including cut-off), and elaborate the flow diagrams with the unit processes. The total system of unit processes involved in the life cycle is called the product system (Guinée 2002). In the next step, data is collected, the flows of emissions, wastes, resources, etc., are gathered and they can be primary data (e.g., measured, etc.) or secondary data given by proper available LCI (Life Cycle Inventory) data sets. There are also other important issues that should be addressed like data quality, data validation and allocation.

In the impact assessment phase, the data from the inventory analysis is converted into potential environmental impacts. It involves several steps: selection of impact categories, classification, characterization, normalization, aggregation and weighting, of which the last three are optional.

In the first step, the impact categories are chosen (ex.: climate change, human toxicity, etc.). Then the different substances/compounds are placed in each category and a value to the indicator is calculated and this corresponds to classification and characterization. In normalization, the magnitude of the indicator relatively to a reference value is calculated and in aggregation, different categories are joined according to, for example, the type of damage. Weighting consists of attributing weights to each category that reflects their relative importance. There are a number of impact assessment methods which are used to calculate impact assessment results (CML 2001, Eco-indicator 99, etc.).

Finally in the interpretation phase, the main conclusions are drawn and recommendations are formulated.

### LCA Application in Soil Contamination

Environmental impacts related to contaminated soil are usually classified in three categories, namely primary impacts, secondary impacts, and tertiary impacts. The primary impacts are related with local toxic impacts associated with contaminated sites. The risks for human health and ecosystems are highly dependent on local characteristics such as specific contaminants, site conditions, and level of exposure of receptors. Due to these characteristics, the assessment of primary impacts are usually based on risk assessment. The protection of soil itself is not yet a priority and major concerns are related to groundwater contamination and human exposure via drinking water. This may be due to the lack of specific legislation on this issue. The secondary impacts consider the local, regional, and global impacts arising from the extraction, materials use, end-of-life stages of all consumables, equipment, and energy used for the remediation. Tertiary impacts deal with environmental impacts associated with the future use of the site. Most LCA studies in this area can be considered as attributional (describes the product life cycle). Goal and scope of those studies is centered in the environmental comparison between remediation technologies and tertiary impacts are usually excluded (Hu et al. 2011; Gallagher et al. 2013). Some works in this area combine risk assessment and life cycle assessment (Lemming et al. 2012).

Most LCA studies considered as the functional unit of the treatment of an amount of soil or groundwater and recent works also follow this pattern (Hu et al. 2011; Gallagher et al. 2013; Cappuyns 2013a) mainly because it fulfills the requirements for being the functional unit for the goal and scope of those studies: clearly defined, measurable, and provides a basis of comparison based on an equivalent (Morais and Delerue-Matos 2010).

In fact LCA studies for soil remediation are done since the late 1990s and since then a wide range of remediation technologies have been studied, such as landfilling, stabilization/solidification, soil washing, incineration, etc. Another important aspect is that existing sustainability assessment in this area uses LCA methodology to compare remediation technologies (Hou and Al-Tabbaa 2014).

### Constraints in LCA Methodology Applied to Soil Contamination

LCA methodology can be a very useful tool in this area since it can help in decision-making processes. However for that same reason, it is important to be aware of constraints and limitations of LCA in what concerns remediation of contaminated soil.

Many LCA studies do not consider in their goal and scope definition primary and tertiary environmental impacts and this is an important issue because there are interrelated decisions concerning the choice of objectives related to physical state of site, the future of site, and remediation technology to be used (Morais and Delerue-Matos 2010).

Another important issue is the product system definition and the choice of processes to be included in system. However, this is a common problem when applying LCA methodology and requires a thorough analysis to avoid a rather arbitrary and subjective procedure especially when the goal of the study is the comparison of alternative products/services.

The lack of inventory data can also be a problem as well as the lack of a relation between inventory data and a specific region which can be a source of uncertainty. The temporal coverage and the time horizon are also a matter of some difficulty. For example, long-term emissions are often excluded but some authors say they can affect the results of LCA studies. The same happens with capital equipment and infrastructure (Morais and Delerue-Matos 2010).

The method and impact categories used also affect the outcome of LCA studies and its choice is very important (Cappuyns 2013b). However there are guidelines to help to make a decision taking in consideration the goal and scope of the study (Guinée 2002).

As mentioned before many LCA studies do not consider primary impacts which can be best assessed by risk assessment since it predicts the local risks. For that reason some authors applied risk assessment in site remediation LCA studies (Lemming et al. 2012; Inoue et al. 2011).

The assessment of tertiary impacts is also very difficult because it needs the specification of future use of land and this will imply the consideration of different scenarios to be analyzed (Cappuyns 2013b).

## Conclusions

Soil contamination is a major problem in many countries and regions around the world, from Africa to Asia and including Europe and other continents. The number of potentially contaminated sites is huge and there is also a diversified range of contaminants (heavy metals, organic compounds, etc.).

Problems associated with soil can be of several types (pollution, sealing, salinization, etc.) but all of them contribute to a loss of quality of soil although the definition of soil quality is not yet consensual.

Like many natural resources, soil is a much requested and valuable resource, from agriculture to urbanization, transport, industry, etc. and its management has not been done in a sustainable way.

Due to this reason, it is important to prevent soil contamination and to remediate contaminated soil in spite of the lack of specific legislation on soil protection.

Remediation of contaminated soil is perceived as a positive action but no action is free of environmental impacts. Nowadays there is a variety of remediation technologies available and it is important to apply methodologies such as LCA that can be useful in decision-making processes since it is important to choose remediation technologies that have low environmental impacts.

LCA has scientific bases however there are some limitations and constraints associated with its application. Taking this in consideration it is important to reduce arbitrariness, subjectivity, and uncertainty in LCA studies.

To complement LCA information it can be necessary to use other methods such as risk assessment and economic and social criteria to reach a sustainable remediation and a sustainable use of soil.

## References

Campbell, A. 2008. Managing Australia's Soils: A policy discussion paper. National Committee on Soil and Terrain.

Cappuyns, V. 2013a. Environmental impacts of soil remediation activities: quantitative and qualitative tools applied on three case studies. J. Cleaner Prod. 52: 145–154.

Cappuyns, V. 2013b. LCA based evaluation of site remediation, Opportunities and Limitations. Green Chemistry Sustainability. Chemistry Today 31: 18–21.

Carroll, D., B. Sleep, M. Krol, H. Boparai and C. Kocur. 2013. Nanoscale zero valent iron and bimetallic particles for contaminated site remediation. Adv. Water Resour. 51: 104–122.

[CCICED] China Council for International Cooperation on Environment and Development 2010. Developing Policies for Soil Environmental Protection in China. Annual General Meeting 2010.

European Commission. 2006. COM (2006) 231 final Communication from the Commission to the Council, the European Parliament, the European Economic and Social Committee and the Committee of the Regions, Thematic Strategy for Soil Protection.

European Parliament and Council. 2013. DECISION No. 1386/2013/EU of the European Parliament and of the Council of 20 November 2013 on a General Union Environment Action Programme to 2020 'Living well, within the limits of our planet'.

[EPA] United States Environmental Protection Agency 1992. Contaminants and Remedial Options at Wood Preserving Sites.

[EPA] United States Environmental Protection Agency 1993. Innovative Site Remediation Technology, Soil washing/Soil Flushing, Volume 3.

[EPA] United States Environmental Protection Agency 1994. *In Situ* Vitrification Treatment.

[EPA] United States Environmental Protection Agency 1995. *In Situ* remediation Technology, Electrokinetics.

[EPA] United States Environmental Protection Agency 1998. Evaluation of Subsurface Engineered Barriers at Waste Sites.

[EPA] United States Environmental Protection Agency 2000. Solidification/Stabilization Use at Superfund Sites.

[EPA] United States Environmental Protection Agency Region 1. 2010. Clean and Green Policy for Contaminated Sites.

[EPA] United States Environmental Protection Agency 2014. http://www.epa.gov/superfund/sites/npl/(accessed on 16 July 2014).

Gallagher, P., S. Spatari and J. Cucura. 2013. Hybrid life cycle assessment comparison of colloidal silica and cement grouted soil barrier remediation technologies. J. Hazard. Mater. 250-251: 421–430.

Guineé, J.B. 2002. Handbook on Life Cycle Assessment—Operational Guide to the ISO Standards. Kluwer Academic Publishers, Dordrecht.

Hou, D. and A. Al-Tabbaa. 2014. Sustainability: A new imperative in contaminated land remediation. Environ. Sci. Policy 39: 25–34.

Hu, X., J. Zhu and Q. Ding. 2011. Environmental life-cycle comparisons of two polychlorinated biphenyl remediation technologies: Incineration and base catalyzed decomposition. J. Hazard. Mater. 191: 258–268.

Inoue, Y. and A. Katayama. 2011. Two scale evaluation of remediation technologies for a contaminated site by applying economic input-output life cycle assessment: Risk-cost, risk-energy consumption and risk-$CO_2$ emission. J. Hazard. Mater. 192: 1234–1242.

ISO. 2006. ISO 14040:2006, Environmental management—Life cycle assessment—Principles and framework.

Liedekerke, M. van, G. Prokop, S. Rabl-Berger, M. Kibblewhite and G. Louwagie. 2014. Progress in the management of Contaminated Sites in Europe. European Commission, EUR 26376—Joint Research Centre—Institute for Environment and Sustainability.

Lemming, G., J.C. Chambon, P.J. Binning and P.L. Bjerg. 2012. Is there an environmental benefit from remediation of a contaminated site? Combined assessments of the risk reduction and life cycle impact of remediation. J. Environ. Manage. 112: 392–403.

Marker, A. 2013. Manual: Revitalização de áreas degradadas e contaminadas (brown fields) na América latina, Projeto Integration, ICLEI.

Morais, S. and C. Delerue-Matos. 2010. A perspective on LCA application in site remediation services: Critical review of challenges. J. Hazard. Mater. 175: 12–22.

Panagos, P., M. van Liedekerke, Y. Yigini and L. Montanarella. 2013. Contaminated sites in Europe: Review of the current situation based on data collected through a European network. J. Environ. Public Health 2013: 11 pages.

[UNCCD] United Nations Convention to Combat Desertification 2012. Zero Net Land Degradation, A Sustainable Development Goal for Rio + 20.

US Department of Energy 2014. http://energy.gov/em/services/site-facility-restoration/soil-groundwater-remediation (accessed on 16 July 2014).

Varanda, M., G. Pinto and F. Martins. 2011. Life cycle analysis of biodiesel production. Fuel Process. Technol. 92: 1087–1094.

# CHAPTER 7

# Nanoremediation with Zero-valent Iron Nanoparticles

*S. Machado*[1] *and J.G. Pacheco*[1,*]

## ABSTRACT

Nanotechnology is becoming more and more important in today's society, being responsible for the development of innovative products and technologies within distinct areas. The use of nanomaterials to deal with environmental contamination begun at the end of the last century, achieving, nowadays, a reasonable state of development. Nanoremediation involves the application of extremely reactive nanomaterials that can act as catalysts, react, or immobilize the contaminants reducing/eliminating the risk to human health or to the ecosystems.

There are a significant number of materials that have been used for nanoremediation, such as metal oxides, enzymes, carbon nanotubes, and other metallic materials, namely zero-valent iron. Nanoparticles of zero-valent iron (nZVI) are the most common and widely used material for environmental remediation. They have shown high efficiency in the treatment of chlorinated compounds (such as trichloroethylene and perchloroethylene) and metals (chromium).

This chapter focuses on the role of nZVI as remediation agent, its chemical and physical properties that make this material so attractive, the main synthesis methods, the state of the art related to laboratorial studies, the applications that have been performed in contaminated sites, and the (eco)toxicity of this material in distinct organisms.

[1] REQUIMTE/LAQV, Instituto Superior de Engenharia do Porto, Instituto Politécnico do Porto, Rua Dr. António Bernardino de Almeida 431, 4200-072 Porto, Portugal.
* Corresponding author: jpgpa@isep.ipp.pt.com

## Introduction

The advances of nanotechnology led to the development of innovative materials and domestic and industrial applications within distinct areas such as medicine, textiles and cosmetics production, military, semiconductors industry, display and optical technologies, food industry, and environmental remediation, namely soil remediation (Karn et al. 2009; Crane and Scott 2012).

The social awareness regarding soil contamination issues is increasing and has led to several efforts that aim to identify, study, and remediate contaminated sites. The lists of confirmed contaminated sites as well as the compounds that are considered contaminants are increasing and demand actions to remediate them in order to prevent and assure human health and preserve soil functions (Zhang 2003). Therefore, and to accomplish this goal, it is essential to have all the tools available, namely analytical methodologies and remediation technologies. Among the most recent technologies, nanoremediation can be highlighted because of its potential to become a cost-effective and versatile remediation technology. Nanoparticles are particles with at least one of its dimensions below 100 nm, presenting specific properties in mechanical strength, electrical conductivity, magnetism, and bio-inhibition (Pan and Xing 2012). These particles have key properties for soil remediation: effective transport through groundwater and porous media, persistence in suspension for long periods of time, and easy deployment in slurry reactors for the remediation of contaminated solid wastes, soils and sediments, and high reactivity for the removal of the contaminant (Zhang 2003; Crane and Scott 2012). However, the possible toxicity of nZVI should be considered in order to act consciously.

Nanoparticles, some of them derived from natural materials, can be used for adsorption and immobilization of heavy metals, such as cadmium (Ghrair et al. 2009), arsenic (An and Zhao 2012), mercury (Xiong et al. 2009), and chromium (He and Zhao 2007). Furthermore, it is known that several contaminants are highly reactive with several nanomaterials, such as nanoscale zeolites, metals and its oxides, carbon nanotubes and fibers, and noble metals and titanium dioxide (Karn et al. 2009). Silver nanoparticles have high reactivity with several aqueous contaminants. However, the cost of this material as well as its environmental toxicity, excludes them from environmental applications (AshaRani et al. 2009). On the other hand, iron nanoparticles (mostly zero-valent) are the most widely studied because of their low cost and high reactivity and adsorption capacity (Zhang 2003). Table 1 presents the nanoparticles that can be used for remediation and the contaminants that can be potentially degraded (Zhang 2003; Theron et al. 2008).

As presented in Table 1, there are several types of nanoparticles with capacity to be used in environmental remediation, being nZVI the one that can degrade a wider range of contaminants. Following this, the focus of this

**Table 1.** Nanomaterials used for environmental remediation and the contaminants potentially remediated.

| Nanomaterial | Contaminants |
|---|---|
| Activated carbon fibers | BTEX and heavy metal ions |
| Bimetallic nanoparticles | PCBs, chlorinated ethenes, chlorinated methanes |
| Carbon nanotubes | BTEX and heavy metal ions |
| Carbon nanotubes functionalized with polymers and iron | p-nitrophenol, benzene, toluene, dimethylbenzene and heavy metal ions |
| Nanocrystalline zeolites | Toluene and nitrogen dioxide |
| Ni/Fe nanoparticles and Pd/Au nanoparticles | PCBs, chlorinated ethenes, chlorinated methanes |
| TiO$_2$ photocatalysts | Heavy metal ions, azo dyes, phenol, aromatic pollutants and toluene |
| Zero-valent iron nanoparticles (nZVI) | Chlorinated methanes, trihalomethanes, chlorinated benzenes, chlorinated ethenes, pesticides, polychlorinated hydrocarbons, organic dyes, heavy metal ions, inorganic ions, chlorinated organic compounds |

chapter will be on nZVI, the iron chemistry, the nZVI synthesis methods and its applications related to environmental remediation and (eco)toxicological issues regarding their use.

Iron is one of the most abundant elements on our planet and exists in two predominant valence states: ferrous (Fe (II)) and ferric (Fe (III)) ions and the relatively water-soluble Fe (II) (ferrous iron) and zero valent (Fe (0)-metallic or elemental). However, this last one, due to its extremely high reactivity, namely with oxygen, is rarely found; as soon as it is formed, it generally reacts with the surrounding environment (Cundy et al. 2008).

The oxidation of zero-valent iron originates several types of iron oxides such as hematite ($\alpha$-Fe$_2$O$_3$), maghemite ($\gamma$-Fe$_2$O$_3$), magnetite (Fe$_3$O$_4$), goethite ($\alpha$-FeO·OH), lepidocrocite ($\gamma$-FeO(OH)), and ferrihydrite (Fe$_2$O$_3$·½H$_2$O). Considering this high tendency to rapidly react with oxygen, it is important to increase the stability of nZVI in order to guarantee better performances in subsoil and in groundwater. Following this, distinct stabilizing agents have been used to passivate the highly reactive surface of the nanoparticles from reacting with the surrounding media, namely through resin and starch (He and Zhao 2005), carboxymethylcellulose, and chitosan (Geng et al. 2009). The capping of the nanomaterials will also have a direct and negative effect on the reactivity of the nanomaterials with the contaminants; however this is compensated by the size of the nanomaterials and the respective gain in surface area (He et al. 2007).

Zero-valent iron is an excellent electron donor (presenting a standard reduction potential ($E^0$) of –0.440 V) as described in the following equation (Matheson and Tratnyek 1994):

$$Fe^0 \rightarrow Fe^{2+} + 2\ e^-$$

This property makes zero-valent iron a very attractive material for environmental remediation. The first and most common application of zero-valent iron as environmental remediation agent (in permeable reactive barriers) was to degrade chlorinated contaminants such as the trichloroethylene (TCE) or perchloroethylene (PCE) (Ibrahem et al. 2012). However, other studies showed that it can be used for a wide range of other contaminants, namely metals (e.g., chromium and arsenic) (Thiruverikatachari et al. 2008; Franco et al. 2009), polychlorinated biphenyls (PCBs) (Chen et al. 2014), and organochlorine pesticides (Huang et al. 2012).

## Synthesis Methods

nZVI can be synthesized through two distinct approaches: top-down and bottom-up. The former includes the utilization of physical and/or chemical methods to reduce the size of or restructure bulk materials to nanoscale while the latter builds up nanomaterials from basic structures, such as atoms or molecules also using physical or chemical methods. Examples of top-down methods are: pulsed laser ablation, milling processes, or electroexplosion (Zhang 2003; Crane and Scott 2012).

The **pulsed laser ablation** has the capacity to produce stable and pure nanoparticles (Fig. 1).

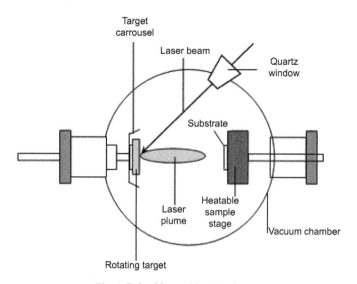

**Fig. 1.** Pulsed laser ablation scheme.

The setup of this method involves a laser pulse beam that operates at a defined repetition rate (e.g., 10 Hz) and fundamental wavelength (Cristoforetti et al. 2012) that focuses on a target surface at a normal incidence by means of a lens that leads to the ablation of the material and nanoparticle production.

The **milling processes** consist of crushing macro- or microparticles to produce nanoparticles and are generally used for metallic or ceramic materials. One example of this type of equipment is the high-energy ball mill that is constituted by one turn disc, two or four bowls and zirconium dioxide milling balls. The turn disc rotates in one direction while the bowls rotate in the opposite direction applying strong centrifugal forces to the milling balls and to the material to be crushed. This way the material is fractured and cold welded under high energy impact inflicted by the mill balls (Tolia et al. 2011).

**Electroexplosion** is based on the application of a very short and very powerful pulse of current to a metallic wire. The temperature of this wire increases to 25000–30000°C, pulverizing it into its atomic constituents. A glowing cloud of plasma is formed and held together by the strong magnetic field that accompanies the pulse. The metal cloud reacts with the noble gas that fills the reactor forming the nZVI. The shape and the sizes of the synthesized nanoparticles depend on several factors, such as the shape and size of the wire, the applied voltage, and the nature of the applied electrical pulse (Schulenburg 2008).

The most common and **traditional bottom-up method** for the production of nZVI consists of the chemical reaction between sodium borohydride and iron (III) in aqueous solution:

$$4Fe^{3+}(aq) + 3BH_4^-(aq) + 9H_2O \rightarrow 4Fe^0(s) + 3H_2BO_3^-(aq) + 12H^+(aq) + 6H_2(g)$$

This method is very fast and simple but it presents several disadvantages such as the toxicity of sodium borohydride, that obliges to complementary actions in order to remove the remaining borohydride from the produced nanoparticles, and the production of gaseous hydrogen during the synthesis, which requires safety precautions (Machado et al. 2013a). In this type of method, the concentration of the reagents, the temperature, the pH of the solution, the sequence in which the source materials are added and mixed are factors that influence the characteristics of the nanoparticles. One limitation of iron nanoparticles is the possibility of agglomeration, which can be avoided by attaching a stabilizer such as a soluble polymer or surfactant onto the nanoparticles.

More recently, a new **green synthesis method** has been developed. In this method sodium borohydride is replaced by natural extracts with high antioxidant capacities (namely polyphenols). The first studies about this subject describe the preparation of nZVI using tea polyphenols (Hoag et al. 2009). The nZVI were prepared within a few minutes at room temperature by mixing a tea extract with a ferric nitrate solution, without any need of adding any surfactants/polymers as capping agent. More recently, similar approaches based on this type of synthesis have been evaluated based on the utilization of leaves from fruit and forest trees (Machado et al. 2013a; Wang et al. 2014) (Fig. 2). This green method presents several advantages, besides the use of natural material extracts instead of sodium borohydride, when compared to the traditional method: it is fast and simple, and the used extract offers capping of the produced nanoparticles hindering their agglomeration tendency and can act as source of nutrients for a complementary biodegradation phase (Machado et al. 2013a).

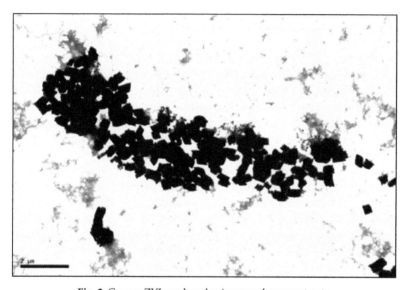

**Fig. 2.** Green nZVI produced using pear leaves extracts.

Another difference between the two bottom-up methods is the amorphous nature that the synthesized "green" nanoparticles present while the nZVI produced using sodium borohydride have Fe crystalline phases for Fe° (Jameia et al. 2013).

The synthesis of nZVI is the key point of this technology because the particle properties, such as size, reactivity, and mobility, are strongly dependent on the manufacturing process.

One important challenge is the need to scale up and focus on the development of economical methods for the production of large quantities of nZVI. These methods should also assure the quality and the control of the particles' properties. The use of sub-products of the food industry with high polyphenol contents for the green synthesis of nZVI could be an answer to these challenges, providing stabilized nanoparticles and large-scale and *in situ* production.

## Applications

### *Laboratorial and Pilot Tests*

The potential of nZVI for environmental remediation led to numerous laboratorial tests concerning the reactivity of these nanoparticles with distinct contaminants, its transport in porous media, and its toxicity in the various environmental compartments in which it can be applied. In this sub-section, the laboratorial studies that brought advances to possible applications of nZVI in soil remediation will be presented.

nZVI has been successfully used to degrade a wide range of organic and inorganic water and soil contaminants, including several chlorinated solvents, pesticides, aromatic nitro compounds, azo dyes, and heavy metals (Zhang et al. 2009; Satapanajaru et al. 2009, 2011). Remediations of halogenated hydrocarbons using nZVI have been particularly well studied and documented, in part because they are some of the most common soil and groundwater contaminants. Recently, efforts have been made in order to extend this technology to other kind of compounds such as, for example, pharmaceuticals (Fang et al. 2011; Machado et al. 2013b) and polycyclic aromatic hydrocarbons (PAHs) (Chang and Kang 2009).

Several works focus on the advantages that could be achieved by capping the nanomaterials and therefore stabilizing the particles' surfaces, enhancing their activity. Reyhanitabar et al. (2012) evaluated the efficiencies of distinct iron particles, including nZVI, stabilized with starch to physically sorb and subsequently reduce Chromium (VI). These nanoparticles were the ones that presented higher performances. Similar conclusions were made by Li et al. (2011); however, in silica, fume was used to support nZVI. Darko-Kagya et al. (2010) evaluated the reactivity of nZVI and aluminum lactate-modified nZVI with 2,4-dinitrotoluene (2,4-DNT) in soils with two different permeabilities (kaolin and field sand). In this work the aluminum lactate cap around the nZVI led to the decrease of the amount of 2,4-DNT removed/degraded after one day of reaction. However, it increased with time; in kaolin soil, after the second day, the degradation/removal of 2,4-DNT was already similar while in field sand this occurred only after 14 days. The decrease of the removal/degradation of the contaminant was due to the formation of a thin film layer around the nZVI leading to a reduction of its surface area and consequently of its reactivity. In another study, Ponder et al. (2000) produced nZVI using the traditional bottom-up method supported in a nonporous, hydrophobic resin, achieving rates of metal (Cr (VI) and Pb) remediation up to 30 times higher for the supported nZVI than for the unsupported ones.

Basnet et al. (2013) studied the influence of capped nanoparticles on their mobility in porous media. Natural and organic macromolecules, such as carboxymethyl cellulose, rhamnolipid biosurfactants, and soy protein, were used to coat palladium-doped zero-valent iron nanoparticles (Pd-nZVI) and the aggregation and transport behavior of these particles were compared to bare Pd-nZVI. The results showed that the coated Pd-nZVI had good colloidal stability and presented enhanced transport and that biosurfactants were the most suitable surface modifiers for field applications. Other authors coated nZVI with calcium hydroxide and observed that this coating led to an effective protection layer for the nZVI, improved its mobility in sand columns and had no negative effect on the reactivity of the nZVI when used for the removal of chromium (VI) from groundwater (Wei and Li 2013).

## Site Remediation

Apart from the know-how that scientific research is supplying, other advances rise from field applications. Since 2000, field demonstrations and land applications have been conducted using nanoremediation to treat soils; in the last years these applications have increased in number and spread through all continents, mostly in North America. In Karn et al. (2009), 44 worldwide sites where nanoremediation has been used were presented: 38 located in North America (two in Canada and the remaining in the United States of America), five in Europe (two in the Czech Republic, one in Germany, Slovakia, and Italy) and one in Asia (Taiwan). However, it should be considered that these numbers correspond exclusively to the reported projects. The geographical distribution of these projects is closely related to the national legislations that are more or less rigid to the application of nanomaterials in soils or in waters.

Considering the type of nanomaterials that were used in the above mentioned projects, nZVI was used as remediation agent in almost two thirds of these projects, and in 6% of the studies emulsified zero-valent iron was used, showing once more the potentialities of this material. The remaining projects used other materials such as capped and uncapped Fe/Pd nanoparticles. Considering the sites remediated using nZVI, the contaminants that were addressed and efficiently removed were: vinyl chloride (VC), dichloroethylene (DCE), trichloroethylene (TCE), perchloroethylene (PCE), polychlorinated biphenyls (PCBs), methylene chloride, 1,2-dichloropropane, 1,2-dichlorethane, chromium (VI), pesticides (dichlorodiphenyldichloroethylene (DDE) and dichlorodiphenyltrichloroethane (DDT)), and perchlorate.

The presented remediation projects generally applied nZVI for short periods of time, from one month to one year, but there are cases where the treatment lasted for more than five years. During this time the contaminants' reductions were generally extremely high (above 90% in several cases where the contaminants were TCE or DCE), however in some cases the removal was lower, 55–60%.

Another study evaluated the performance (deliverability and reactivity) of carboxymethyl cellulose stabilized nZVI in a sandy aquifer of Alabama contaminated with chlorinated solvents. The results showed that the nanoparticles were delivered 10 ft downgradient from the injection well and that the contaminant reduction efficiencies were up to 94% in the first week of injections. The application of stabilized nZVI and the consequent reduction of contamination in the soil boosted biological processes that led to a long-term reduction of the concentration of the chlorinated contaminants in the soil and in the groundwater (Man and Dongye 2013). However, this work points some technical limitations that this technology presents, namely the impact that soil sorption has on the rate and extend of the degradation. This is one of the operational problems that the use of nZVI has; other authors have reported different restrictions related with the nZVI mobility and interaction with soil constituents. For example, the tendency that nZVI have to agglomerate and form larger particles that can block the soil pores, limiting its transport, and hindering them to reach the contaminated zones of the soil and therefore perform the remediation (Cameselle et al. 2013).

## Toxicity of nZVI

The use of nanomaterials for environmental remediation is increasing significantly as well as the concern about their possible toxicity. In the United Kingdom the use of nanomaterials for soil remediation is not allowed because of the uncertainties related to their toxicity. This position is based on the precautionary principle, which is a fundamental principle that supports the European Union's health and environmental policy. All these concerns are related to the lack of sufficient studies that clearly and scientifically describe the effects of nanomaterials in the ecosystems and in the distinct organisms. Peralta-Videa et al. (2011) compiled the information of several works published in the biennium 2008 and 2010 concerning the use of nanomaterials and concluded that there are still uncertainties on the nanomaterials toxicity and reinforced the need for more information on the best way to handle those materials in order to preserve the environment and human health. In this section the state of the art concerning the toxicity of nZVI in different stages of the food chain (bacteria, aquatic invertebrates, and terrestrial organisms) are presented.

**Bacteria and algae** are considered the bottom of the aquatic food chain and the food of several aqueous crustaceans (e.g., Daphnia). *Escherichia coli, Klebsiella planticola,* or *Bacillus nealsonii* are some of the strains that are commonly used for toxicity tests. Toxicity tests performed with some of these strains and nZVI showed distinct results. According to Barnes et al. (2010), the presence of nZVI hindered the biodegradation of TCE and for concentrations of nZVI higher than 0.3 g $L^{-1}$, the biodegradation was interrupted. However in Kirschling et al. (2010), no toxic effect of nZVI was perceived on the abundance of bacteria in the soil, in fact the bacterial populations increased when polyaspartate coated nZVI was applied to soils indicating that the use of coatings can decrease the nanoparticles toxicity. However, this study did not present long term results, in order to evaluate the nZVI impact after the biodegradation of the organic capping by the endogenous microorganisms.

**Aquatic invertebrates** are commonly used for toxicity tests because they are considered the final recipients of most of the most common contaminants. Among the distinct invertebrates, *Daphnia magna* is probably the most used for toxicity tests (Sanchez et al. 2011). According to Baun et al. (2008), tests with these organisms are recommended in order to evaluate and understand the toxicity of nanomaterials in aquatic environments. The knowledge of the impact of nZVI on aquatic invertebrates is still very scarce but all of them concluded that the nZVI negatively and significantly affects the *Daphnia magna* even causing their death, indicating that the use of this nanomaterial for treatment of aqueous systems should always be performed in accordance with the protection of the ecosystems (Sanchez et al. 2011). In Marsalek et al. (2012), nZVI was used to destroy and prevent the formation of cyanobacterial water blooms; the nanomaterial has an $EC_{50}$ of 50 mg $L^{-1}$ against cyanobacteria, much lower than against *Daphnia magna* (1000 mg $L^{-1}$). In another work, commercial coated and uncoated nZVI (Nanofer 25S and Nanofer STAR) were tested also with *Daphnia magna* and showed to be toxic for this organism, drastically decreasing the growth rates and increasing mortality (Keller et al. 2012).

The use of nZVI for soil remediation is increasing and therefore requires information about the impact of this material on **terrestrial organisms**, such as earthworms. These organisms are usually used for ecotoxicity tests, mostly in situations where the contamination has already occurred (Sanchez et al. 2011). However the knowledge concerning the impact of nZVI on earthworms is extremely scarce: as far as it is known there is only one work published. In this work, El-Temsah and Joner (2012) observed a negative impact of nZVI on the reproduction, weight changes, and mortality rates of two species of earthworms, *Eiseniafetida* and *Lumbricusrubellus*.

## Acknowledgements

This work received financial support from the European Union (FEDER funds through COMPETE) and National Funds (FCT, Fundaçãopara a Ciência e a Tecnologia) through projects Pest-C/EQB/LA0006/2013, PTDC/AAG-TEC/2692/2012, and PTDC/AAG-TEC/4403/2012. To all financing sources the authors are greatly indebted. J.G. Pacheco also thanks FCT for financial support through his post-doc grant SFRH/BPD/101419/2014.

## References

An, B. and D. Zhao. 2012. Immobilization of As(III) in soil and groundwater using a new class of polysaccharide stabilized Fe-Mn oxide nanoparticles. J. Hazard. Mater. 211: 332–341.

AshaRani, P.V., G.L.K. Mun, M.P. Hande and S. Valiyaveettil. 2009. Cytotoxicity and genotoxicity of silver nanoparticles in human cells. ACS Nano 3: 279–290.

Barnes, R.J., O. Riba, M.N. Gardner, A.C. Singer, S.A. Jackman and I.P. Thompson. 2010. Inhibition of biological TCE and sulphate reduction in the presence of iron nanoparticles. Chemosphere 80: 554–562.

Basnet, M., S. Ghoshal and N. Tufenkji. 2013. Rhamnolipid biosurfactant and soy protein act as effective stabilizers in the aggregation and transport of palladium-doped zerovalent iron nanoparticles in saturated porous media. Environ. Sci. Technol. 47: 13355–13364.

Baun, A., N.B. Hartmann, K. Grieger and K.O. Kusk. 2008. Ecotoxicity of engineered nanoparticles to aquatic invertebrates: A brief review and recommendations for future toxicity testing. Ecotoxicology 17: 387–395.

Cameselle, C., K.R. Reddy, K. Darko-Kagya and A. Khodadoust. 2013. Effect of dispersant on transport of nanoscale iron particles in soils: Zeta potential measurements and column experiments. J. Environ. Eng.-ASCE 139: 23–33.

Chang, M.C. and H.Y. Kang. 2009. Remediation of pyrene-contaminated soil by synthesized nanoscale zero-valent iron particles. J. Environ. Sci. Health, Part A: Toxic/Hazard. Subst. Environ. Eng. 44: 576–582.

Chen, X., X.Y. Yao, C.N. Yu, X.M. Su, C.F. Shen, C. Chen, R.L. Huang and X.H. Xu. 2014. Hydrodechlorination of polychlorinated biphenyls in contaminated soil from an e-waste recycling area, using nanoscale zerovalent iron and Pd/Fe bimetallic nanoparticles. Environ. Sci. Pollut. Res. 21: 5201–5210.

Crane, R.A. and T.B. Scott. 2012. Nanoscale zero-valent iron: Future prospects for an emerging water treatment technology. J. Hazard. Mater. 211: 112–125.

Cristoforetti, G., E. Pitzalis, R. Spiniello, R. Ishak, F. Giammanco, M. Muniz-Miranda and S. Caporali. 2012. Physico-chemical properties of Pd nanoparticles produced by Pulsed Laser Ablation in different organic solvents. Appl. Surf. Sci. 258: 3289–3297.

Cundy, A.B., L. Hopkinson and R.L.D. Whitby. 2008. Use of iron-based technologies in contaminated land and groundwater remediation: A review. Sci. Total Environ. 400: 42–51.

Darko-Kagya, K., A.P. Khodadoust and K.R. Reddy. 2010. Reactivity of lactate-modified nanoscale iron particles with 2,4-dinitrotoluene in soils. J. Hazard. Mater. 182: 177–183.

El-Temsah, Y.S. and E.J. Joner. 2012. Ecotoxicological effects on earthworms of fresh and aged nano-sized zero-valent iron (nZVI) in soil. Chemosphere 89: 76–82.

Fang, Z., J. Chen, X. Qiu, X. Qiu, W. Cheng and L. Zhu. 2011. Effective removal of antibiotic metronidazole from water by nanoscale zero-valent iron particles. Desalination 268: 60–67.

Franco, D.V., L.M. Da Silva and W.F. Jardim. 2009. Reduction of hexavalent chromium in soil and ground water using zero-valent iron under batch and semi-batch conditions. Water Air Soil Pollut. 197: 49–60.

Geng, B., Z. Jin, T. Li and X. Qi. 2009. Preparation of chitosan-stabilized Fe-0 nanoparticles for removal of hexavalent chromium in water. Sci. Total Environ. 407: 4994–5000.

Ghrair, A.M., J. Ingwersen and T. Streck. 2009. Nanoparticulate zeolitic tuff for immobilizing heavy metals in soil: Preparation and characterization. Water Air Soil Pollut. 203: 155–168.

He, F. and D.Y. Zhao. 2005. Preparation and characterization of a new class of starch-stabilized bimetallic nanoparticles for degradation of chlorinated hydrocarbons in water. Environ. Sci. Technol. 39: 3314–3320.

He, F. and D. Zhao. 2007. Manipulating the size and dispersibility of zerovalent iron nanoparticles by use of carboxymethyl cellulose stabilizers. Environ. Sci. Technol. 41: 6216–6221.

He, F., D. Zhao, J. Liu and C.B. Roberts. 2007. Stabilization of Fe-Pd nanoparticles with sodium carboxymethyl cellulose for enhanced transport and dechlorination of trichloroethylene in soil and groundwater. Ind. Eng. Chem. Res. 46: 29–34.

Hoag, G.E., J.B. Collins, J.L. Holcomb, J.R. Hoag, M.N. Nadagouda and R.S. Varma. 2009. Degradation of bromothymol blue by 'greener' nano-scale zero-valent iron synthesized using tea polyphenols. J. Mater. Chem. 19: 8671–8677.

Huang, J., X. Cong and Q. Gu. 2012. Factors influencing the dechlorination of organo-chlorine pesticides in soils of a contaminated site by zero-valent iron. Disaster Adv. 5: 105–108.

Ibrahem, A.K., T. Abdel Moghny, Y.M. Mustafa, N.E. Maysour, F. Mohamed Saad El Din El Dars and R. Farouk Hassan. 2012. Degradation of trichloroethylene contaminated soil by zero-valent iron nanoparticles. ISRN Soil Sci. 2012, 9.

Jameia, M.R., M.R. Khosravi and B. Anvaripour. 2013. Degradation of oil from soil using nano zero valent iron. Sci. Int. 25: 863–867.

Karn, B., T. Kuiken and M. Otto. 2009. Nanotechnology and *in situ* remediation: A review of the benefits and potential risks. Environ. Health Perspect. 117: 1813–1831.

Keller, A.A., K. Garner, R.J. Miller and H.S. Lenihan. 2012. Toxicity of nano-zero valent iron to freshwater and marine organisms. PloS One 7: e43983.

Kirschling, T.L., K.B. Gregory, E.G. Minkley, Jr., G.V. Lowry and R.D. Tilton. 2010. Impact of nanoscale zero valent iron on geochemistry and microbial populations in trichloroethylene contaminated aquifer materials. Environ. Sci. Technol. 44: 3474–3480.

Li, Y., Z. Jin, T. Li and S. Li. 2011. Removal of hexavalent chromium in soil and groundwater by supported nano zero-valent iron on silica fume. Water Sci. Technol. 63: 2781–2787.

Machado, S., S.L. Pinto, J.P. Grosso, H.P.A. Nouws, J.T. Albergaria and C. Delerue-Matos. 2013a. Green production of zero-valent iron nanoparticles using tree leaf extracts. Sci. Total Environ. 445: 1–8.

Machado, S., W. Stawiński, P. Slonina, A.R. Pinto, J.P. Grosso, H.P.A. Nouws, J.T. Albergaria and C. Delerue-Matos. 2013b. Application of green zero-valent iron nanoparticles to the remediation of soils contaminated with ibuprofen, Sci. Total Environ. 461-462: 323–329.

Man, Z. and Z. Dongye. 2013. *In situ* dechlorination in soil and groundwater using stabilized zero-valent iron nanoparticles: Some field experience on effectiveness and limitations. pp. 79–96. *In*: S. Ahuja and K. Hristovski (eds.). Novel Solutions to Water Pollution. American Chemical Society, Washington, DC, USA.

Marsalek, B., D. Jancula, E. Marsalkova, M. Mashlan, K. Safarova, J. Tucek and R. Zboril. 2012. Multimodal action and selective toxicity of zerovalent iron nanoparticles against cyanobacteria. Environ. Sci. Technol. 46: 2316–2323.

Matheson, L.J. and P.G. Tratnyek. 1994. Reductive dehalogenation of chlorinated methanes by iron metal. Environ. Sci. Technol. 28: 2045–2053.

Pan, B. and B.S. Xing. 2012. Applications and implications of manufactured nanoparticles in soils: a review. Eur. J. Soil Sci. 63: 437–456.

Peralta-Videa, J.R., L. Zhao, M.L. Lopez-Moreno, G. de la Rosa, J. Hong and J.L. Gardea-Torresdey. 2011. Nanomaterials and the environment: A review for the biennium 2008–2010. J. Hazard. Mater. 186: 1–15.

Ponder, S.M., J.G. Darab and T.E. Mallouk. 2000. Remediation of Cr(VI) and Pb(II) aqueous solutions using supported, nanoscale zero-valent iron. Environ. Sci. Technol. 34: 2564–2569.

Reyhanitabar, A., L. Alidokht, A.R. Khataee and S. Oustan. 2012. Application of stabilized Fe0 nanoparticles for remediation of Cr(VI)-spiked soil. Eur. J. Soil Sci. 63: 724–732.

Sanchez, A., S. Recillas, X. Font, E. Casals, E. Gonzalez and V. Puntes. 2011. Ecotoxicity of, and remediation with, engineered inorganic nanoparticles in the environment. Trends Anal. Chem. 30: 507–516.

Satapanajaru, T., S. Onanong, S.D. Comfort, D.D. Snow, D.A. Cassada and C. Harris. 2009. Remediating dinoseb-contaminated soil with zerovalent iron. J. Hazard. Mater. 168: 930–937.

Satapanajaru, T., C. Chompuchan, P. Suntornchot and P. Pengthamkeerati. 2011. Enhancing decolorization of Reactive Black 5 and Reactive Red 198 during nano zerovalent iron treatment. Desalination 266: 218–230.

Schulenburg, M. 2008. Nanoparticles—small things, big effects, Opportunities and risks. Bundesministerium für Bildung und Forschung (BMBF), Referat "Nanomaterialien; Neue Werkstoffe", Federal Ministry of Education and Research, Bonn, Germany.

Theron, J., J.A. Walker and T.E. Cloete. 2008. Nanotechnology and water treatment: Applications and emerging opportunities. Crit. Rev. Microbiol. 34: 43–69.

Thiruverikatachari, R., S. Vigneswaran and R. Naidu. 2008. Permeable reactive barrier for groundwater remediation. J. Ind. Eng. Chem. 14: 145–156.

Tolia, J.V., M. Chakraborty and Z.V.P. Murthy. 2011. Mechanochemical synthesis and characterization of Group II–VI semiconductor nanoparticles. Part. Sci. Technol. 30: 533–542.

Wang, T., X.Y. Jin, Z.L. Chen, M. Megharaj and R. Naidu. 2014. Green synthesis of Fe nanoparticles using eucalyptus leaf extracts for treatment of eutrophic wastewater. Sci. Total Environ. 466: 210–213.

Wei, C.J. and X.Y. Li. 2013. Surface coating with Ca(OH)(2) for improvement of the transport of nanoscale zero-valent iron (nZVI) in porous media. Water Sci. Technol. 68: 2287–2293.

Xiong, Z., F. He, D. Zhao and M.O. Barnett. 2009. Immobilization of mercury in sediment using stabilized iron sulfide nanoparticles. Water Res. 43: 5171–5179.

Zhang, W.X. 2003. Nanoscale iron particles for environmental remediation: An overview. J. Nanopart. Res. 5: 323–332.

Zhang, X., Y.-M. Lin and Z.-L. Chen. 2009. 2,4,6-Trinitrotoluene reduction kinetics in aqueous solution using nanoscale zero-valent iron. J. Hazard. Mater. 165: 923–927.

# Managing Contaminated Groundwater
## Novel Strategies and Solutions Applied in The Netherlands

*Hans Slenders,*[1,*] *Rachelle Verburg,*[1] *Arnold Pors*[1] and
*Anouk van Maaren*[1]

## ABSTRACT

The Netherlands has a long history of soil and groundwater remediation.
Often clean up proved to be hindered by technical, juridical and financial
barriers and there obviously was a need for more cost-efficient and effective
approaches. As a result, we have seen a shift from a policy strongly focussed
on soil protection towards a risk-based sustainable use of soil and subsurface.
As the remedial operation progressed, there was also a shift from mainly
soil remediation towards groundwater challenges; overlapping plumes
and interaction with other subsurface use. In this chapter two innovative
developments are introduced that contribute to a sustainable subsurface use;
area wide groundwater management, and the combination of remediation
and Aquifer Thermal Energy Storage. These developments have created a
real breakthrough and are illustrated with 3 full scale examples in Eindhoven,
Utrecht and Het Gooi.

---
[1] ARCADIS, P.O. Box 1018, 5200 BA, The Netherlands.
* Corresponding author: Hans.Slenders@arcadis.nl

## The Main Challenge: Groundwater Plumes

The Netherlands was one of the first countries to undertake remediation of contaminated sites. Since the early 1980s, the investigation and remediation of contaminated sites has been high on the agenda, and the recent Midterm Review (Uitvoeringsprogramma Bodemconvenant 2013) revealed that the end of the process of managing historically contaminated sites is in sight. The review gave a comprehensive overview of the remaining contaminated sites that need remediation. Out of a total of 1539 remaining sites, at 1177 sites the risk as a result of spreading contamination plumes impacting groundwater quality is the dominant factor (whereas the risk for humans or ecology near the surface is a priority at 362 sites).

The Netherlands are a delta region, and the soil consists mainly of sandy and clayey deposits. Since the industrial revolution, pollutants have seeped into the subsurface and caused plumes of contamination within aquifers and gradually dispersed over larger areas. As a result, contaminated groundwater lies beneath most urbanised and industrial areas. This threatens the quality of surrounding groundwater wells or surface waters and is potentially harmful for humans or the environment.

Decades of experience with the remediation of polluted groundwater has shown that the clean-up is hindered by cost and technical challenges. Often plumes originating from different sites and different problem-owners mingle, and a separate, standalone approach per plume is not feasible due to the way

Fig. 1. Contaminated groundwater areas in The Netherlands.

plumes interact. Liability issues then hamper an effective approach. Traditional solutions are therefore not viable for a large portion of contaminated plumes, making it difficult to improve groundwater quality or to protect vulnerable areas. Moreover, the interference with other subsurface activities such as groundwater extractions for building pits or cooling/heating purposes pose extra challenges.

A parallel development over the last decade has been the growing demand for sustainable energy, and the use of subsurface water to create heat cold storage (HCS) or Aquifer Thermal Energy Systems (ATES) for the buildings on the surface. More than 1,800 ATES systems have been installed in the Netherlands, with plans to significantly increase their numbers in the next decade. These initiatives are mostly being employed in urban areas, where there is a need for offices and other buildings, but where historically contamination is also present. Under these circumstances, because moving and spreading of contaminants is prohibited, the effective use of groundwater for ATES is often unattainable.

Developments over the last 5–10 years have supplied a set of solutions that can be used to meet all these challenges. The solutions are no longer just technical in nature. They are a combination of policy renewal, technical development, and last but not least, organisational skills to overcome liability issues and to stimulate cooperation. At the start of this chapter, the major building blocks are discussed:

- A policy shift from sectoral protection to sustainable subsurface management;
- The organisation of area-wide groundwater management;
- Aquifer Thermal Energy Storage, principles and possibilities.

At the end of this chapter, real life cases are presented that illustrate that all these building blocks are needed to come to a robust solution. Finally, the Netherlands certainly is not unique with regard to the groundwater problems or intended groundwater use. It is anticipated that populated areas around the globe will be encountering similar challenges. Whether the solutions described in this chapter will work in other settings still has to be proven. But we are convinced that it is worth trying, and with the right planning and recognition of the value of the solution, it could become an important way forward in many cases.

## Developments

### *A Policy Shift*

#### *Policy shifts in soil protection legislation*

Since soil protection legislation was first introduced in the Netherlands it has undergone several changes. Until the mid-seventies there was little or no awareness that pollutants could have an unacceptable and adverse effect on our soil and groundwater. Thereafter four stages of policy maturity can be distinguished. At first it was assumed that the number of contaminated sites was limited and that full restoration was feasible (multi-functionality). Very soon it became apparent that site specific exemptions were needed at sites where containment seemed the only solution. In the early 2000s, a risk based approach was integrated in legislation. And the ongoing policy evolution led to the understanding that our subsurface should be fit for use, with a quality of soil and groundwater with a level of risk for man, plant, and animal as low as "reasonably" possible. In this case, "reasonableness" or cost-effectiveness can be expressed as an optimal balance between the benefits and impacts of remedial activities (Ministerie van I&M 2013), which is only little less than the concept of sustainable remediation (CL:AIRE 2010; NICOLE 2010). In Fig. 2 the evolution in soil policy is depicted as an underground journey. Currently the train is approaching the last stop.

Again the Netherlands are not unique, the developments in many European countries are alike.

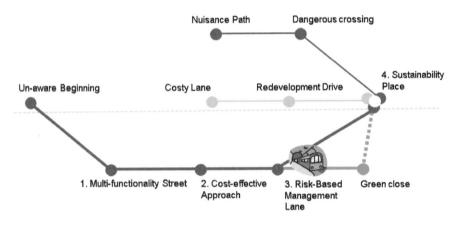

**Fig. 2.** The Underground Journey.

*Policy shifts in groundwater management*

In recent years the soil legislation developments also resulted in a more pragmatic, risk-based approach to groundwater quality management, which on the one hand aims to overcome the challenge of overlapping contaminant plumes with multiple "owners", and on the other hand tries to benefit from groundwater use as a resource. The current groundwater policy is based on three goals:

1) **Protect:** protect the environment and contain contaminated groundwater to prevent further spreading of contaminants;
2) **Use:** benefit from groundwater's potential, e.g., as a sustainable form of energy supply to heat or cool buildings;

3) **Improve:** enhance the quality of groundwater, by active measures, or by optimising the conditions for the natural breakdown of contaminants over time.

Since 2012 the Dutch Law on Soil and Groundwater Protection contains an article (Wet Bodembescherming 2013, art. 55c-i WBB) that enables Area Wide Groundwater Management (AWGM, see below). In this article, provisions have been made such that an authority can take over the liability for a cluster of plumes, whereas the responsibility for the remediation of (shallow) source zones remains with the site owner. The goals for the contiguous contaminated groundwater areas are as mentioned above: protect, use, and improve, but actual improvement may be slow and take time.

For management of groundwater, the Water Framework Directive (European Parliament and Council 2000) and the Groundwater Daughter Directive (European Parliament and Council 2006) are also important. They set objectives for pollution trends and to prevent and limit the input of pollutants to groundwater. The current interpretation however is that they are only applicable for contaminant plume sizes that are significant at the scale of groundwater bodies, and only a very limited portion of the historical contaminant plumes is this big. Nevertheless, for AWGM, also for smaller areas, the upward groundwater quality trend is adopted as a standard requirement. This is to avoid discussions as to what level a contaminant situation is, or is not significant at a groundwater body scale (In the Netherlands the Kempen area and the Rotterdam Harbour area are assumed to be significant).

### Area-Wide Groundwater Management

In the previous paragraph on policy developments area-wide groundwater management (AWGM) was mentioned. But why move to AWGM, and what is it exactly?

### Reasons to consider AWGM

According to Dutch law, AWGM can be applied if: "... *contaminant plumes in groundwater ... are mixed ... or... if a separate management approach is not possible without influencing another plume...*" (Wet Bodembescherming 2013, art. 55c). This can also be interpreted as meaning that an individual approach simply is not possible. But there are more reasons:

- At greater depths, a distinction between plumes often is no longer possible, and it is not viable to appoint one responsible party for a certain contaminated volume;
- Without some sort of organisation or juridical basis the required cooperation between plume owners in most cases will not be successful;
- Solution for stagnation. Without the framework of AWGM the presence of contaminants often hinders redevelopment. Building pit extractions

are not possible or expensive and need to be accompanied by a remedial plan and permits. The largest part of the contiguously contaminated areas are found in city centres, which are characterised by a high level of activity and dynamic spatial developments;
- Barriers for sustainable subsurface use. Under traditional conditions ATES systems are not permitted in contaminated groundwater, because they (can) lead to uncontrollable dispersion;
- Technical challenges; and
- Excessive cost.

### Juridical situation

The juridical situation around groundwater contamination is complex. In the Netherlands groundwater is a "res nullius", which means that it belongs to nobody, until it is extracted. Also, there is no direct right to be able to abstract clean groundwater. If a groundwater plume has been legally recognised before an intention to use the groundwater as a resource is known, under normal circumstances it is difficult to hold the owner of a plume liable, even if the contamination causes extra cost to third parties when they want to extract or use groundwater for another purpose. All are seen as subsurface use, and the one that comes first, normally has first right. It is also not permitted to cause extra expansion of a plume as a result of groundwater extraction.

### Main characteristics of AWGM

An important aspect of AWGM is that shallow sources of contaminants in the soil on one hand and the deeper contaminated groundwater areas on the other hand, are treated separately, and most likely by different stakeholders. The area-wide approach is only applied to the deeper groundwater contamination. The reason for this is that the risks at ground level belong to the shallow sources (< 5 m) and remain the responsibility of the site owner. At this shallow level, the different contaminant composition and extents can still be differentiated and the relation between the contamination and the original source locations is still evident, making a case-based, location-specific approach viable. The site owner is responsible for the implementation of the required remedial actions in the source zone.

By contrast, in deeper groundwater, the connection between the original source and the mixed plumes is often not evident, as distinct plumes can no longer be distinguished. Furthermore, the plume often extends underneath the properties of other parties, and the risks of the contaminants are completely different. A zone is defined as that which comprises the contaminated area, and by disregarding the individual cases there is room for additional mixing caused by subsurface use, e.g., ATES or building pit extractions. This is allowed within the boundaries of the management zone, as long as the groundwater quality improves in the longer term. A zoned approach to groundwater management

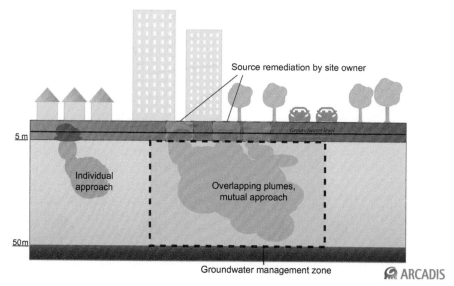

**Fig. 3.** Disconnection of sources and plumes.

therefore allows for functional ground water use, while permitting natural attenuation processes to contain the extent of the contamination and help it to reduce in concentration over time. The clear choice of a monitoring line or Plain of Compliance (PoC) and a set of intervention measures are needed to protect the surrounding areas.

The AWGM approach usually is not limited by a timescale, and therefore the management of the area cannot be in the hands of a commercial organization, but will be put in the hands of an administrative body. Handing over the management of a contaminant plume from a private problem owner towards an area manager also implies the transfer of responsibilities and liabilities and thus a transfer sum. The area manager will only take over possible liabilities if a financial arrangement is made.

The left side shows that in a case-based approach, ATES implementation is not really possible without moving contaminants. There are also many monitoring schemes around the plumes, which interfere with each other. In a zoned approach on the right-hand side there may only be one monitoring scheme (or plain of compliance), which is the circle surrounding the area. The sustainable use of the groundwater is possible as long as the surrounding uncontaminated groundwater is protected against contaminants.

AWGM satisfies the objectives of the general groundwater policy by:

- **Improvement**: source removal and (enhanced) natural attenuation;
- **Protection**: receptors at surface level through source remediation and surroundings through monitoring;
- **Use**: extractions and ATES systems are allowed (aiding promotion of redevelopment).

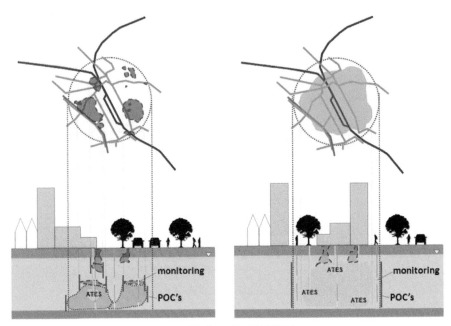

**Fig. 4.** The benefits of AWGM.

### *Aquifer Thermal Energy Storage or Heat Cold Storage*

An aquifer thermal energy system uses groundwater and a heat exchanger to cool buildings, and via a heat pump, to warm them. For this purpose twin wells have to be created, a combination of an extraction well and an infiltration well, usually in an anaerobic aquifer. The background temperature of groundwater (typically 12 degrees C in the Netherlands) is perfectly suited for use in heat pumps. By reversing the flow between the two wells, the flow direction can be changed in cold and warm seasons. Colder and warmer zones can thus be created in groundwater. Aquifers are well suited for the storage of warm or cold water, as the subsurface has good thermodynamic and insulating properties. Temperatures are well maintained. Previously warmed water can be used in winter for the heating of buildings. The surplus of cold in winter is then stored in the aquifer, and used in summer for the cooling of buildings. By storing heat and cold, the energy efficiency of the heat pumps is further enhanced.

Given the soil profile in the Netherlands, shallow ATES are applicable nearly everywhere, and the advantages of ATES are apparent:

- A significant reduction in energy consumption;
- As a result a reduction in fossil fuel consumption;
- Lowering of carbon dioxide emissions.

These advantages mean that a significant cost reduction can be achieved and the return on investment time is short (only several years).

In general there are three types of ATES systems that move groundwater. While there are also closed loop systems, or geothermal systems, they are not the subject here. The three types of open groundwater systems are:

- Heat Cold Storage systems;
- Twin wells (heat-cold zones in the same soil layer or aquifer);
- Monowell (heat-cold separated vertically in different layers);
- Recirculation system, the flow direction is not changed, and there is no storage of heat or cold. Groundwater is extracted at the same temperature continuously. The above ground heat pump is designed to produce heat or cold with a constant starting temperature.

In all three systems groundwater is pumped out of or into aquifers, resulting in a significant change in the natural groundwater flow. In case the groundwater is contaminated or there is a contaminant plume nearby, the risk of additional dispersion of contaminants is big (BOEG 2010). However, this can be mitigated through careful system design.

## Cases

### Sanergy in Eindhoven; ATES as a Remediation Tool

#### Terra incognita to be explored

As a result of decades of industrial activity, soil and groundwater at a site in Eindhoven are contaminated with "chlorinated solvents" (halogenated hydrocarbons) to a depth of 60 to 70 m below ground surface (bgs). During planning of the redevelopment of the area into mixed development for housing and offices, very ambitious criteria were formulated. A sustainable society was one of the missions, and the intention was introduced to combine groundwater remediation with groundwater as an energy source. ATES systems have the promise of reducing at least 50% of greenhouse gas emissions. But in 2004, the idea of combining these two concepts was completely new, and it was not clear if it was feasible from a technical and legislative viewpoint.

#### Challenges

The idea did entail a couple of challenges to overcome. Aquifer Thermal Energy Systems use large flows to transport heat and cold. This is contrary to the remediation goal of controlling and reversing the dispersion of contaminants. This hurdle can only be overcome if the required large flows are used to contain the contaminants as part of an overall ATES system. The hydrogeological design has to be adjusted to this aim.

Within an ATES system, the flow in the extraction wells is determined by the energy demands of the buildings. However, remediation requires that continuous flows are used to contain contaminated groundwater and to manipulate the groundwater flow field. Reconciling the demand for energy with the need for containment is a complex but essential task. It requires another conceptual way of thinking, and a smart design of the system; coupling of wells; management of pumps, sensors, the in-house installations.

Finally there are also organisational challenges. These are not only in the cooperation between experts of different backgrounds, for example, hydrologists with installation expertise etc. but often also obtaining agreement between different authorities. In Eindhoven, remediation work falls under the jurisdiction of the municipality, while the groundwater energy is the responsibility of the province.

### Drafting of the solution

In a traditional ATES-system with a warm and a cold storage zone it is not possible to contain the groundwater flow. It is also difficult to anticipate fluctuations in groundwater flow directions. All the challenges and difficulties were overcome by reverting to a groundwater recirculation system, a system where the infiltration wells are surrounded by a "cage" of extraction wells (see Fig. 5). The infiltration wells alternate between infiltration of heated or cooled water, dependent upon the need for energy. The system can be compared with a huge *in situ* remediation system used to alter groundwater quality, that has a continuous flow direction and can be regarded as a controlled hydrogeological subsystem.

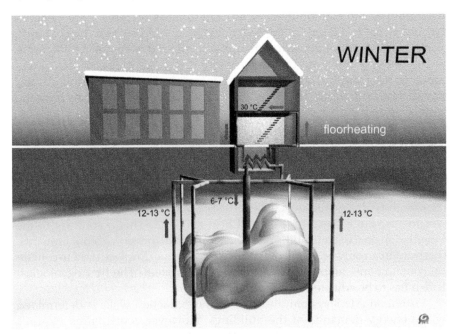

**Fig. 5.** The concept of Sanergy.

## Experiences after four years of operation

The system was designed and approved in 2008, and installed in 2009. In operation since 2010, the flow pumped gradually increased until the current is 100 m³/hr, which is needed to create a containment effect. The system is designed for an average volume of 300 m³/hr, but redevelopment at the site slowed down and as a result the need for the extraction of heat and cold out of the groundwater also decreased. The groundwater volume between the extraction wells has been flushed 2–3 times since the commencement. Initially the increase in groundwater flow within the system led to an increase of DOC (dissolved organic carbon) and consequently an increase of natural degradation of the contaminants (chlorinated solvents, Meer met Bodemenergie 2011). More recently, monitoring data indicate that the groundwater quality in the system is more homogenised, with suboptimal conditions for microbial degradation. Within the circumference of the extractions wells concentrations slowly equalized as a result of mixing. In Fig. 6 hydrogen concentrations are illustrated over time. It is commonly assumed that degradation starts at levels between 1 and 2 nM H₂. The concentrations now seem to converge to levels around 1 nM.

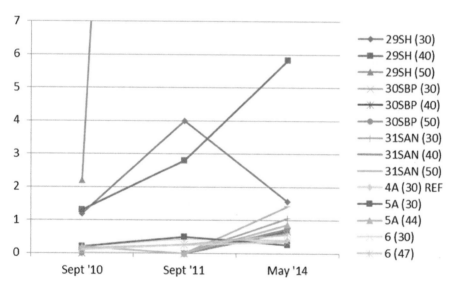

**Fig. 6.** Hydrogen concentrations (nM).

But apart from the natural degradation, monitoring of contaminants and a parallel groundwater modelling study confirm that the Sanergy system has a strong containing effect and that the downstream concentrations will decrease. The source flux from the site has decreased, and attenuation downstream will lead to lower concentrations.

## Groundwater use in an Area-wide Approach, a Cost-effective Solution in Utrecht

### Stagnating development

The inner city of Utrecht has been undergoing major renovation work in recent years, across a sizable area, which has led to a number of activities involving the subsurface. These include groundwater extractions for building excavations, and the development of underground infrastructures and ATES systems to cool and heat new offices and buildings.

As a city with a long history, Utrecht has been home to many industrial activities in the past, and as a result, its groundwater is extensively contaminated. There is an estimated 50 million $m^3$ of groundwater in the aquifer where ATES systems have been planned, an aquifer comprising some 29 sources and plumes of chlorinated solvents and other chemicals, plumes that cannot be distinguished as separate entities any more.

Some 20 ATES systems are being employed or are planned, ranging from 5 to 50 m below ground surface, across an area of roughly 6 $km^2$. This has presented a major obstacle, given the usual ban on the activities which could result in further dispersion of contaminated water. Due to the interaction between the plumes and the difficult setting in an historic city, remediation work would be far too expensive and time consuming to implement effectively on a case-by-case basis.

The breakthrough came early in 2010, with the approval of a new Remedial Action Plan for the area (Dols 2009). The municipality worked hard on the required policy interpretations and initiated the actual planning and design. The Remedial Plan allowed for an area-wide approach; a defined system area, or management zone, within which groundwater contaminants would be allowed to move and mix, to allow for the establishment of ATES- systems.

### Modelling of groundwater flow and contaminant behavior

The municipality wanted to know what the effects of the ATES systems would be on the contaminant situation, and it was decided to build an RT3D solute transport model that could simulate flow, and the degradation and retardation of the chlorinated solvents. A monitoring network was installed and some data gaps in knowledge about the groundwater quality were addressed. Subsequently the model was populated with data relating to the contaminants and microbial degradation rates (half-life times). The half-life times were estimated by correlating the ratios between the breakdown products of per- and trichloroethene over three different plumes. In Fig. 7 the predicted future trends in vinyl chloride concentrations over a 30 year period are illustrated. In the centre of the figure (within the black contour), the attenuating effect of the ATES systems over time is apparent. It is also obvious that the total

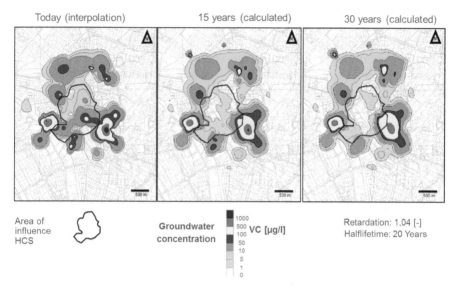

**Fig. 7.** Development of vinyl chloride concentrations over time.

contaminated area does not really grow in size significantly, although the degradation rates used in the modelling exercise are very low and likely to be conservative.

There are also strong indications that the overall quality of the groundwater will improve over time because of the increased contact between contaminants, nutrients and naturally occurring bacteria, as the groundwater mixes through pumping. This effect was observed in Eindhoven at the Sanergy site.

On the Utrecht site, even using very conservative degradation constants, the total amount of chlorinated solvents is expected to decrease from approximately 6,000 kg to 4,000 kg over a 30-year period; a significant improvement in groundwater quality.

## Huge reduction of emissions and financial costs

It is estimated that the Aquifer Thermal Energy Systems will lead to a 30–50% reduction in $CO_2$ emissions and non-renewable sources for energy, which will be even greater if the reductions achieved through the avoidance of remediation works were also taken into account. From a cost perspective, it was estimated in Utrecht that the zoned, area-wide approach would cost €15–€20 million, versus €500 million for the same area to undergo traditional remediation. The Utrecht Remedial Action Plan was a crucial step forward that is helping improve groundwater quality, while also protecting the surrounding "clean" areas. Since 2010 the plan has been updated once again, because the contaminated area proved to be wider than previously anticipated. The reason for moving towards an AWGM plan was underlined even more. The effective subsurface use of the groundwater in Utrecht has found to be particularly important for facilitating ongoing spatial developments, and last but not least, if applied correctly, should be a valuable tool to aid improvement of groundwater quality. No wonder Utrecht calls this solution "The bio-washing machine".

### Achieving Control in the Gooi area

*An enormous sandpit with contaminant plumes*

The area "Het Gooi" has a surface of approximately 200 km$^2$ and is located in the centre of the Netherlands. The area comprises seven municipalities and four groundwater protection areas for drinking water. Historical industrial activities in this area have led to around 100 contaminant plumes in groundwater, varying in size from several metres to several kilometres. The area is characterized by a soil profile of highly permeable sands from ground surface to as much as 200 m deep. The largest plumes have already reached the groundwater protection areas, and are threatening surface water quality—water that is used for recreational purposes. In the urbanised areas the contaminated plumes are hampering the implementation of ATES and building pit extractions. However, the threats on the drinking water quality are the main issue. Chlorinated solvents from chemical laundries and electronic industries caused serious DNAPL migration (dense non-aqueous phase liquids) and the former wastewater infiltration ponds from the Municipality have facilitated deep migration of various contaminants since the mid thirties of the 20th century. These contaminants are now found in the monitoring and interception wells of the Laren drinking water station. The exact extent of the plumes was not known. In general though, the area can be regarded as a closed hydrogeological system. Infiltrating rainwater will end up in either the drinking water extractions or the surrounding surface waters and polders. In these soils, contaminant dispersion is rapid and deep as a result of the DNAPL and forced infiltration.

*Cooperation is needed*

Full delineation of all the plumes is technically and financially too big a challenge. However, there was a need for effective management of the contaminants, with a better control and view on the risks for the receptors and possibilities for subsurface use. In addition to the technical challenges, it required that several municipalities, the water authorities, drinking water companies, and the Province work together, with clear lines of communication and a transparent process. The Province managed to get an agreement in place between all of the parties and initiated a successful cooperation approach.

### Source-path-receptor linkage for a complete area

The first step was an extensive inventory of all contaminant sources and related data, and subsequent mapping. The second was a complete inventory of vulnerable receptor areas which included surface and swimming water, polders, and drinking water stations. A calibrated groundwater model was used to simulate flowlines between the potential sources and receptors. Finally, a risk-based approach and monitoring strategy was applied. This strategy developed in line with guidance 17 of the Groundwater Daughter Directive (European Commission 2007), which roughly means that there is a monitoring scheme surrounding the contaminated areas, and a monitoring scheme as an early warning system around the vulnerable receptors (protection areas).

### A management plan for the area

A framework plan and six plans for separate Management sub areas (see Fig. 8) were developed for the area: three monitoring plans to safeguard the drinking water stations, two plans for contiguously contaminated areas, and a plan for isolated cases. These plans combine to form a robust strategy to enable the continued use of groundwater for drinking water, while removing the obstacles to other subsurface use. For the first time the extent of the problem became clear, and the first and biggest gain was obvious; no longer was the full area of Het Gooi considered contaminated, but only the old part of Hilversum, two smaller areas in Naarden and Huizen, and around 10 isolated contaminated sites. This meant that more than 80% of the area was considered "not contaminated".

The management plan for the sub area of Hilversum, with many contaminant plumes, contains a monitoring scheme at the downstream side, safeguarding further plume expansion, and controls on subsurface activities. Within the area, contaminants below a depth of 5 m are allowed to mix. In all sub areas and sites the remediation of the sources of contamination remains the responsibility of the site owner, but the responsibility and liability for the plumes can be handed over to a government owned organisation, which takes over the management of the groundwater.

The strategy for the Laren drinking water station was new. The contaminants are already within the groundwater protection area and actually threaten drinking water quality. Groundwater is already being treated before it enters the distribution network. As a result, a robust early warning monitoring system was needed to respond to concentration fluctuations and a further decrease in groundwater quality. Wells are usually monitored at locations where groundwater has 10 to 20 years flow time left before reaching the extraction wells, but in this case monitoring wells have been planned at only two to three years of flow time, giving companies ample time to continue to guarantee safe provision of clean drinking water.

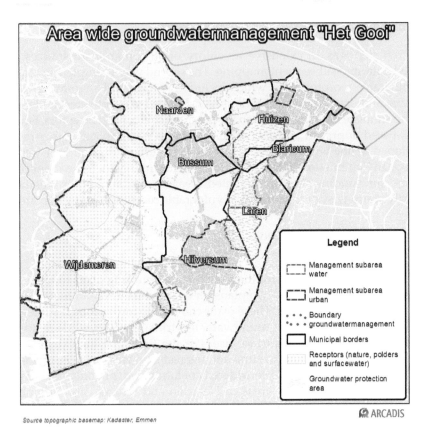

**Fig. 8.** The Gooi area with Management sub areas and receptors.

The framework plan for 'Het Gooi' makes the complex contaminant situation perfectly tangible and reinforces the effective protection of drinking water plants. Given the large volume of contaminated groundwater, the betterment of groundwater quality will take some time, but active source removal, the interception wells near the pumping station, and natural attenuation will in the end lead to an improvement of this part of the groundwater body.

In October 2013 the plan was approved by the Mayors and deputies of the Municipalities and Province of North-Holland, and the directors of the water companies. They regard this approach as a great leap forward. Indeed, the area and plan are unique in their size and the sustainability of approach.

## Concluding: A New Approach for a Sustainable Future

Governments and companies face severe bottlenecks and stagnation in tackling large scale contaminated groundwater areas. The contaminants entail risks, and may delay spatial developments and the application of beneficial subsurface

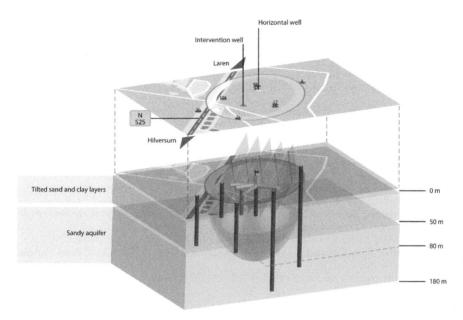

**Fig. 9.** The Laren drinking water extraction (horizontal wells in yellow) and the monitoring wells (purple and pink) on the surrounding ellipse.

activities, such as groundwater energy systems. In some instances, it is local legislation that was needed to manage the further spread of contaminants which is hampering efforts to remediate contaminated groundwater in a sustainable way.

At present, the Netherlands is one of the front runners in this field, where a more effective and pragmatic approach is being applied, but international interest has begun to grow. The projects in Utrecht and Gooi illustrate how a risk-based approach to groundwater quality management in area-wide contaminated areas enables us to protect human health and improve groundwater quality while harvesting the ecological benefits of renewable energy.

Hans Slenders MSc is a Principal Consultant Soil and Groundwater
ARCADIS the Netherlands

## Acknowledgements

The author would like to acknowledge all the parties involved in the projects mentioned: Philips Environment and Safety, Volker Wessels DEC, Brabant Water, Vitens, PWN, Municipality of Utrecht, Province of Northern Holland.

# References

[BOEG] Bodemenergie en Ground water verontreiniging 2010. A guideline for dealing with ATES in contaminated groundwater. SKB and Ministry of the Environment.

[CL:AIRE] Contaminated Land: Applications in Real Environments 2010. A Framework for assessing the sustainability of soil and groundwater remediation, SURF-UK. CL:AIRE, London, UK.

Dols, P. 2009. Saneringsplan Ondergrond Utrecht Gefaseerde gebiedsgerichte aanpak (Remedial plan subsurface Utrecht, phased AWGM), ARCADIS Nederland BV, Hoofddorp, The Netherlands.

European Commission 2007. Guidance Document No. 17, Guidance on preventing or limiting direct and indirect inputs in the context of the groundwater directive 2006/118/EC.

European Parliament and Council 2000. Water Framework Directive 2000/60/EC.

European Parliament and Council 2006. Groundwater Daughter Directive 2006/118/EC.

Meer met Bodemenergie 2011. Effecten van WKO op verontreinigingen: metingen op verontreinigde locaties en labtesten (Effects of ATES on contaminants, measurements at contaminated sites and laboratory tests).

Ministerie van I&M 2013. Circulaire Bodemsanering 2013 (Circular Soil Remediation 2013).

[NICOLE] Network for Industrially Contaminated Land in Europe 2010. NICOLE Road Map for Sustainable Remediation.

Uitvoeringsprogramma Bodemconvenant 2013. Midterm review 2013 Bodemconvenant, Eindversie.

Wet Bodembescherming 2013 (Soil Protection Act 2013).

# Biological Techniques to Remediate Petroleum Hydrocarbons in Contaminated Environments

*Nazaré Couto*[1,*] *and F. Javier García-Frutos*[2,†]

## ABSTRACT

Contamination with petroleum hydrocarbons (PHC) is a worldwide problem with environmental implications. Mineral oil, polycyclic aromatic hydrocarbons, and aromatic hydrocarbons are in the top-4 soil contaminants present in European solid matrices.

Biological remediation treatments are cost-effective and restore soil structure, being a "green" alternative to remediate PHC contaminated sites. The presence of persistent compounds is an additional threat to soil. This chapter aims to overview the potential of biological remediation techniques such as natural attenuation, biostimulation, bioaugmentation, and rhizoremediation to restore these contaminated environments. The synergistic effect with other techniques (e.g., surfactant amendment) is also explored aiming to increase the remediation potential.

## Background

Human activities have led to the deposition of contaminants in soils with severe effects to the environment. Additionally, many of them are persistent

[1] CENSE, Departamento de Ciências e Engenharia do Ambiente, Faculdade de Ciências e Tecnologia, Universidade Nova de Lisboa, 2829-516 Caparica, Portugal.
[2] CIEMAT, Avda. Complutense 40, 28040 Madrid, Spain.
* Corresponding author: md.couto@fct.unl.pt
† The editors condole his passing away. The memories of his life and work will remain forever in the minds and hearts of those who knew him.

and represent a serious threat to the ecosystems. The search for remediation options to promote environmental restoration is of extreme importance.

Two hundred years of industrialization reflect situations of environmental contamination in Europe. In the context of the European Soil Data Centre (ESDAC), the Joint Research Centre (JRC) and the European Environmental Agency (EEA) agreed to harmonize European data related with soil (Liedekerke et al. 2014). The "Progress in the management of Contaminated Sites in Europe" report (Liedekerke et al. 2014) summarizes information on the management of contaminated sites obtained from National Reference Centres for Soil from countries that belong to the European Environment Information and Observation Network (EIONET). From the 36 members of the network, 27 participated in the 2011–2012 campaign for data collection. A tentative extrapolation to all Europe estimates 2.5 million of potentially contaminated sites, of which about 14% are expected to be contaminated and likely to require remediation (Liedekerke et al. 2014). Industrial and commercial activities together with waste disposal represent almost two thirds of local soil contamination (Liedekerke et al. 2014). The contaminants affecting the solid matrix (soil, sludge, and sediment) are (by decreasing percentage): heavy metals (35%), mineral oil (24%), polycyclic aromatic hydrocarbons (PAH; 11%), aromatic hydrocarbons (benzene, toluene, ethylbenzene, and xylene—BTEX; 10%), chlorinated hydrocarbons (8%), phenols (1%), and cyanides (1%); the remaining contaminants ("others") represent 10% of solid matrix contamination (Liedekerke et al. 2014).

Soil excavation and disposal is applied in approximately 30% of sites and *in situ* and *ex situ* treatment options are applied with similar frequencies (Liedekerke et al. 2014). The report shows that the most frequently applied treatment techniques to remediate the contaminated soils include "*ex situ* treatment (excavation)" dominant in Finland and France, "other soil treatment (including excavation and disposal)" dominant in Norway, Slovakia, and the United Kingdom, and "*in situ* physical/chemical treatment" dominant in Malta and the Netherlands. The maximum percentage of "*in situ* biological treatment" was reported in the Netherlands (20%), followed by Lithuania, Belgium (Flanders), France, and Italy (in all cases, in the range between 10 and 20%). The maximum percentage of "*in situ* biological treatment (including excavation)" was reported in Estonia (more than 95%), followed by Hungary (approx. 30%), Lithuania (approx. 25%), Belgium (Flanders) and Slovakia (approx. 20%) and, finally, Austria, Finland, Italy, the Netherlands, and Norway (approx. 10%).

## Biological Remediation Technologies

The use of biological techniques to remediate contaminated matrices encloses several advantages such as cost effectiveness, rehabilitation of soil structure, high public acceptance, and the possibility to apply over large areas.

When using biological techniques microorganisms and/or plants have a decisive role in decreasing contaminant levels. In the case of bioremediation, microorganisms will use the organic contaminants as a carbon source which can end in complete mineralization of the contaminant, formation of compounds with lower molecular weights, or transformation in other compounds, for example, more polar with same the same carbon (C) number (Aldrett et al. 1997).

Contaminant-degrading bacteria can be found, virtually, in all soils (Gerhardt et al. 2009). But before an oil spill these microorganisms are generally present in a low amount compared with the total number of microorganisms. When a contamination of petroleum hydrocarbons (PHC) occurs, specific degraders start to proliferate as they use PHC as a carbon source (Bento et al. 2005; Couto et al. 2012). The action of indigenous microorganisms in removing contamination is usually called natural attenuation, i.e., the "self-capacity" of the system to decrease contaminant levels without a specific applied treatment.

The aging of contamination is a very important parameter regarding the effectiveness of the remediation process. After a spill, the more volatile and soluble fractions will volatilize, leach, biotransform, or chemically react whereas the other PHC fractions will increase their relative ratio in the soil matrix (Banks et al. 2003). The hydrophobicity of PHC will lead to a strong sorption to the soil matrix thus reducing availability to microorganisms and limiting the rate of biodegradation (Di Gennaro et al. 2008; Frutos et al. 2011). The slow biodegradation may be attributed to the slow desorption or dissolution rates of hydrophobic organic compounds. In general, soils collected in the field with weathered contamination present a lower remedial behavior than soils spiked in laboratorial conditions (Huang et al. 2005).

The nature of contaminant and the site characteristics also influence the effectiveness of biological techniques. Factors such as level of nutrients, pH, moisture content, presence of microbial toxicants, and characteristics of indigenous microorganisms, amongst others should be adequate to perform an effective remediation scheme. When these conditions are not "optimal", other technologies may help to increase remediation efficiency. Independent of the applied technique, during and at the end of a remediation scheme, the soil toxicity should be taken into account (e.g., presence of new (toxic) metabolites) (Frutos et al. 2012).

Biostimulation is the adjustment of environmental conditions aiming to stimulate indigenous microorganisms to metabolize contaminants. The addition of nutrients may be used in soils with low nutrient content and/or in the presence of a high input of PHC in the environment (due to an accidental spill), which unbalances the carbon:nitrogen:phosphorus (C/N/P) ratio. Literature reports the positive effect of nutrient addition in enhancing the degradation capabilities of indigenous communities, even in polar conditions (Couto et al. 2014; Delille et al. 2004). Either organic or inorganic amendments may be used. Agamuthu et al. (2013) tested the potential of organic wastes to enhance biodegradation of soil contaminated with lubricant oil during 98 days. The addition of sewage sludge and cow dung [10% (w/w)] increased degradation (cow dung: 94%; sewage sludge: 82%) compared to the control. Poultry droppings improved remediation of total petroleum hydrocarbons (TPH) and PAH in soils in a six-week period (Ezenne et al. 2014). Biostimulation of indigenous microorganisms with a lignocellulosic substrate enhanced the remediation of high recalcitrant PAH fraction (high molecular weight) and residual fraction of TPH, the remaining contamination of a soil that had been subject to a pilot-scale biopile treatment for 180 days (Lladó et al. 2013). With this study, biostimulation showed potential to enhance the remediation of some of the most recalcitrant PHC fractions. Biostimulation with kitchen waste and low-level nutrient preparation were used in a biopile scheme and resulted in the removal of more than 80% of initial soil contamination (14 g/kg) thus being important in removing polar components from the contaminated matrix (Liu et al. 2011).

Bioaugmentation relies on the addition of a specific strain or consortia of microorganisms able to degrade the specific contaminant or group of contaminants (Agarry et al. 2010; Bento et al. 2005). This strategy may be used in different situations, for example when the number of specific PHC degraders is low, in areas requiring longer acclimation periods, and in the presence of recalcitrant and/or non-bioavailable compounds. One possibility to perform bioaugmentation is to isolate indigenous microorganisms from the contaminated soil, increase their number under laboratory conditions and re-inoculate the pre-adapted bacterial strains (Mrozik and Piotrowska-Seget 2010). Bioaugmentation is frequently combined with biostimulation. Suja et al. (2014) reported a positive effect in the presence of nutrient addition and bioaugmentation from the local microbial consortia. The author also reported that enhanced air stimulation should be coupled to this remediation scheme to increase remediation of bottom soil in column tests. Steliga et al. (2012) reported that indigenous bacteria and fungi inoculation were effective in improving the remediation of weathered drill wastes highly contaminated with TPH and BTEX, applying the technique in the last part of a three-phase remediation scheme. Another option to perform bioaugmentation is the isolation of specific PHC degraders from a contaminated matrix that is not the specific soil to remediate. The exogenous microorganisms may be greatly different from those existent in the target contaminated soil (Bento et al. 2005) and will have to face new biotic and abiotic conditions that may affect the outcome of the bioaugmentation scheme. For these reasons, allochtonous bioaugmentation may face limitations, namely at field scale. Lladó et al. (2013) reported that exogenous mycoaugmentation was limited by native soil microbiota. Winquist et al. (2014) inoculated a soil contaminated with PAH with a specific fungal inoculum, at lab and field scale. In the lab, the inoculum had a positive effect, especially in high molecular weight PAH. In the field, the bioremediation after three months was similar with or without inoculum addition (94%). Liu et al. (2011) introduced, in a PHC contaminated soil, a bacterial consortium with five strains isolated from different environments and capable of degrading diesel and fuel oil. The biopreparation was applied in a biopile and increased the removal of aromatic compounds. The author reported that the bacterial communities were able to remove most of the saturated and partially aromatic hydrocarbons and, after, the fungal community was responsible for the degradation of polar PHC composing of recalcitrant metabolites.

Couto et al. (2010) evaluated the potential of natural attenuation, bioaugmentation with a commercial product and biostimulation in the remediation of recent (turbine oil) and weathered (crude oil) contamination. The studies were carried out at mesocosm scale. At the end of nine months, higher remediation was observed at top soil (0–5 cm) comparing with that observed at the bottom layers (5–20 cm). Bioaugmentation together with nutrient and surfactant amendments increased remediation at the top soil layer (approximately 20% more than natural attenuation) compared with natural attenuation. Bioaugmentation together with nutrient amendment did not significantly enhance TPH degradation compared to natural attenuation suggesting that the presence of physiologically adapted native microorganisms in the soil was determinant for the remediation of TPH.

Bioventing is an *in situ* option (Magalhães et al. 2009) based on the improved soil oxygenation aiming to stimulate degradation performed by indigenous microorganisms (Österreicher-Cunha et al. 2004). Frutos et al. (2010) used an artificially phenanthrene-contaminated soil (1000 mg/kg soil) to study the evolution and kinetics of contaminant removal by bioventing over seven months. In the best conditions, 60% of water holding capacity (WHC) and a C/N/P ratio of 100:20:1, more than 93% (74 mg/kg) of phenanthrene was remediated. The degradation rate varied between 8.8 mg/kg/day in the first month and 4.5 mg/kg/day at the end of the seven months. The group also performed ecotoxicity tests to assess if the reduction in contaminant concentration indicated a decrease of soil toxicity. In fact, the presence of toxic intermediate metabolites or changes in contaminant bioavailability may increase soil toxicity during bioremediation processes (Al-Mutairi et al. 2008). Also, the soil amendments to enhance remediation may increase soil toxicity. Frutos et al. (2010) reported that the toxicity of ammonium (added as a nutrient) was higher than the toxicity of phenanthrene to plants and aquatic organisms. The ammonium toxicity was not only higher when the nutrient was applied (in the beginning of the study) but remained high even after seven months (Frutos et al. 2010). The results of this study corroborate the importance of incorporating an ecotoxicity assessment to estimate the risk of ecological receptors.

In landfarming the contaminated soil is mixed with soil amendments such as bulking agents and nutrients, incorporated into soil that is uncontaminated, and periodically tilled for aeration. Contaminants are degraded, transformed, and immobilized by microbiological processes and oxidation. Soil conditions such as moisture content and frequency of aeration may be controlled to optimize the rate of contaminant degradation.

Frutos et al. (2012) compared bioremediation and landfarming (specific C/N/P ratio and humidity, and periodic aeration) to treat sludge samples (TPH concentration higher than 2200 mg/kg) collected from a soil washing plant. The studies were carried out on laboratorial and field scales. At the lab scale, 57% of TPH were bioremediated in 28 days, whereas landfarming resulted in 85% of reduction at the end of 180 days. In the pilot trial, bioremediation led to 65% of TPH decrease in 28 days whereas landfarming led to a 42% decrease in 90 days. At the end of the remediation process, a decrease in ecotoxicity was observed in the sludge (Frutos et al. 2012).

In a soil remediation scheme, surfactants may increase desorption and mass transfer from soil (Frutos et al. 2011). The nonpolar contaminants may have two different routes: accumulation in the centre of micelles (when the applied concentration is equal or above critical micellar concentration, CMC) or binding to the hydrophobic groups of surfactants (when the applied concentration is below CMC). A surfactant used below CMC has shown effectiveness in mobilizing PAH from contaminated soil from a former gaswork facility, demonstrating the potential to enhance bioremediation by increasing the bioavailability of the contaminant (Frutos et al. 2011). The positive effect of surfactants on degradation was also reported elsewhere (Cheng et al. 2008; Haigh 1996) but results are still contradictory with reports stating their ineffectiveness or inhibition of degradation (Gao et al. 2007; Haigh 1996; Lladó et al. 2013).

Phytoremediation may have a very important role in the remediation of PHC. Amongst the different phytoremediation techniques, the effect of rhizoremediation will be explored. Vegetated soil presents, generally, a higher number of microorganisms than bulk soil (Glick 2010) as plants exude organic substances (organic acids, simple sugars, amino acids, aliphatics and phenolics, alcohols, proteins) into soil which spur microorganisms growth in the rhizosphere zone. This fact will favor the degradation of contaminants. Plants may also increase the chemical extractability and bioavailability of initially unextractable molecules (Liste and Prutz 2006), an advantage in cases of weathered contamination. Different plant species were identified as growing in PHC contaminated sites and the levels of PHC after a remediation scheme were generally higher in unvegetated than in vegetated soil (Couto et al. 2011, 2012; Mohsenzadeh et al. 2010; Peng et al. 2009). Petroleum resistant plants may have root associated fungal strains that may be useful in the bioremediation of contaminated PHC sites (Mohsenzadeh et al. 2010). The number of root-associated fungi species varies according to plant species and there is a higher variation of strains in plants growing in polluted areas than in uncontaminated sites (Mohsenzadeh et al. 2010). The simultaneous application of plant and its root-associated fungal strains was proved to be more effective than their separated application (Mohsenzadeh et al. 2010).

Couto et al. (2012) tested in a field study (refinery environment) two salt marsh plants, *Scirpus maritimus* and *Juncus maritimus* (alone and in association) aiming to improve the removal of recent and weathered PHC contamination.

After 24 months of exposure, *S. maritimus* was efficient in improving PHC remediation, an opposite pattern than presented by *J. maritimus* and respective association. Between the studied soil depths, the one with higher root density (between 5 and 10 cm) presented the highest remediation efficiency (15% of enhancement). In the same layer, the simultaneous addition of surfactant and bioaugmentation (commercial product) increased the remediation performance by 28%. At the microcosm scale, Couto et al. (2011) tested the same salt marsh species (*S. maritimus*, *J. maritimus* and association) and also *Halimione portulacoides* to remediate a soil with weathered contamination (crude oil) and a mixture of weathered and recent contamination (crude oil and turbine oil). The study lasted seven months. In the control (natural attenuation) 3% of old contamination and 42% of mixture of contamination were removed. The most effective plant species in decreasing PHC levels was *S. maritimus* removing 13% and 79% of old and mixture of contamination, respectively. *H. portulacoides* also showed potential to improve PHC remediation although at a lower extent than *S. maritimus* (10% of removal of weathered contamination; 64% of removal of mixed contamination). The presence of *J. maritimus* or of the association resulted in the least effective treatments (removals between 0 and 2% for weathered contamination and 44 and 48% for mixed contamination— similar values to those obtained in natural attenuation).

In some cases, the combination of biological approaches with physico-chemical techniques may be beneficial to overcome some limitations associated with biological remediation techniques. For example, literature reports the beneficial coupling of bioremediation and an electrokinetic process to remediate TPH in soil (Guo 2014). The "hybrid" technique enhanced degradation effectiveness. The microbial diversity decreased in the presence of the electric field but soil pH and temperature did not change (Guo 2014). The advantages of coupled techniques include decreasing levels of recalcitrant contaminants or increased speed of decontamination.

## Conclusions

The pattern of remediation practices is changing due to the increase in regulatory control of landfill operations (and associated costs) and to the development of *ex situ* and *in situ* remediation techniques (Liedekerke et al. 2014).

Biological remediation is an efficient tool to rehabilitate soils contaminated with PHC. It is cost-effective and presents a "green alternative". The performance of biological remediation is intrinsically dependent on the surrounding environment and its effectiveness depends on several factors (e.g., molecular weight and structure of organic contaminant, soil system conditions including nutrients, pH, temperature, and water).

Natural attenuation may be sufficient to restore the ecosystem to the pre-contaminated conditions, without forced human intervention. But constraints associated with scarce or absent contaminant bioavailability, extremes of toxicity, extensive remediation periods, etc., may limit its success. Recent studies have shown the effectiveness of other biological remediation approaches such as biostimulation, bioaugmentation, and rhizoremediation (sometimes also applied with specific "co-adjuvants" such as surfactants). In fact, by means of an equilibrium between soil nutrients the activity of indigenous microorganisms may be stimulated and/or by means of soil (re)inoculation the level of specific PHC degraders may increase. Also, the synergistic action between plants and microorganisms may improve bioremediation in the root zone.

To be a widely used approach, studies applying biological techniques should prove their effectiveness either in laboratory or in field conditions. However, this may be a challenge due to environmental/field constraints. As so, efforts should be put on increased number of integrated and systematic field studies which will allow a larger scale application of biological techniques for environmental remediation.

## Acknowledgements

Nazaré Couto acknowledges Fundação para a Ciência e a Tecnologia for her Post-Doc fellowship (SFRH/BPD/81122/2011).

## References

Agamuthu, P., Y.S. Tan and S.H. Fauziah. 2013. Bioremediation of hydrocarbon contaminated soil using selected organic wastes. Proc. Environ. Sci. 18: 694–702.

Agarry, S.E., C.N. Owabor and R.O. Yusuf. 2010. Studies on biodegradation of kerosene in soil under different bioremediation strategies. Bioremed. J. 14: 135–141.

Al-Mutairi, N., A. Bufarsan and F. Al-Rukaibi. 2008. Ecorisk evaluation and treatability potential of soils contaminated with petroleum hydrocarbon-based fuels. Chemosphere 74: 142–148.

Aldrett, S., J.S. Bonner, M.A. Mills, R.L. Autenrieth and F.L. Stephens. 1997. Microbial degradation of crude oil in marine environments tested in a flask experiment. Water Res. 31: 2840–2848.

Banks, M., P. Schwab, B. Liu, P. Kulakow, J. Smith and R. Kim. 2003. The effect of plants on the degradation and toxicity of petroleum contaminants in soil: A field assessment. Adv. Biochem. Eng. Biotechnol. 78: 75–96.

Bento, F.M., F.A.O. Camargo, B.C. Okeke and W.T. Frankenberger. 2005. Comparative bioremediation of soils contaminated with diesel oil by natural attenuation, biostimulation and bioaugmentation. Biores. Technol. 96: 1049–1055.

Cheng, K.Y., K.M. Lai and J.W.C. Wong. 2008. Effects of pig manure compost and nonionic-surfactant Tween 80 on phenanthrene and pyrene removal from soil vegetated with Agropyron elongatum. Chemosphere 73: 791–797.

Couto, M.N., E. Monteiro and M.T. Vasconcelos. 2010. Mesocosm trials of bioremediation of contaminated soil of a petroleum refinery: Comparison of natural attenuation, biostimulation and bioaugmentation. Environ. Sci. Pollut. Res. Int. 17: 1339–1346.

Couto, M.N., M.C. Basto and M.T. Vasconcelos. 2011. Suitability of different salt marsh plants for petroleum hydrocarbons remediation. Chemosphere 84: 1052–1057.

Couto, M.N., M.C. Basto and M.T. Vasconcelos. 2012. Suitability of *Scirpus maritimus* for petroleum hydrocarbons remediation in a refinery environment. Environ. Sci. Pollut. Res. Int. 19: 86–95.

Couto, N., J. Fritt-Rasmussen, P.E. Jensen, M. Højrup, A.P. Rodrigo and A.B. Ribeiro. 2014. Suitability of oil bioremediation in an Artic soil using surplus heating from an incineration facility. Environ. Sci. Pollut. Res. Int. 21: 6221–6227.

Delille, D., F. Coulon and E. Pelletier. 2004. Biostimulation of natural microbial assemblages in oil-amended vegetated and desert sub-antarctic soils. Microb. Ecol. 47: 407–415.

Di Gennaro, P., A. Franzetti, G. Bestetti, M. Lasagni, D. Pitea and E. Collina. 2008. Slurry phase bioremediation of PAHs in industrial landfill samples at laboratory scale. Waste Manage. 28: 1338–1345.

Ezenne, G.I., O.A. Nwoke, D.E. Ezikpe, S.E. Obalum and B.O. Ugwuishiwu. 2014. Use of poultry droppings for remediation of crude-oil-polluted soils: Effects of application rate on total and poly-aromatic hydrocarbon concentrations. Int. Biodeter. Biodegr. 92: 57–65.

Frutos, F.J.G., O. Escolano, S. García, M. Babín and M.D. Fernández. 2010. Bioventing remediation and ecotoxicity evaluation of phenanthrene-contaminated soil. J. Hazard.Mater. 183: 806–813.

Frutos, F.J.G., O. Escolano, S. García and G.A. Ivey. 2011. Mobilization assessment and possibility of increased availability of PAHs in contaminated soil using column tests. Soil Sed. Contam. 20: 581–591.

Frutos, F.J.G., R. Péreza, O. Escolano, A. Rubio, A. Gimeno, M.D. Fernández, G. Carbonellc, C. Peruchad and J. Lagunad. 2012. Remediation trials for hydrocarbon-contaminated sludge from a soil washing process: Evaluation of bioremediation technologies. J. Hazard. Mater. 199-200: 262–271.

Gao, Y.-Z., W.-T. Ling, L.-Z. Zhu, B.-W. Zhao and Q.-S. Zheng. 2007. Surfactant-enhanced phytoremediation of soils contaminated with hydrophobic organic contaminants: Potential and assessment. Pedosphere 17: 409–418.

Gerhardt, K.E., X.-D. Huang, B.R. Glick and B.M. Greenberg. 2009. Phytoremediation and rhizoremediation of organic soil contaminants: potential and challenges. Plant Sci. 176: 20–30.

Glick, B.R. 2010. Using soil bacteria to facilitate phytoremediation. Biotechnol. Adv. 28: 367–374.

Guo, S., R. Fan, T. Li, N. Hartog, F. Li and X. Yang. 2014. Synergistic effects of bioremediation and electrokinetics in the remediation of petroleum-contaminated soil. Chemosphere 109: 226–233.

Haigh, S.D. 1996. A review of the interaction of surfactants with organic contaminants in soil. Sci. Total Environ. 185: 161–170.

Huang, X.-D., Y. El-Alawi, J. Gurska, B.R. Glick and B.M.A. Greenberg. 2005. A multiprocess phytoremediation system for decontamination of persistent total petroleum hydrocarbons (TPHs) from soils. Microchem. J. 81: 139–147.

Liedekerke, M. van, G. Prokop, S. Rabl-Berger, M. Kibblewhite and G. Louwagie. 2014. Progress in the Management of Contaminated Sites in Europe, JRC Reference Reports, EUR 26376—Joint Research Centre—Institute for Environment and Sustainability, Luxembourg: Publications Office of the European Union (pp. 72).

Liste, H.-H. and I. Prutz. 2006. Plant performance, dioxygenase-expressing rhizosphere bacteria, and biodegradation of weathered hydrocarbons in contaminated soil. Chemosphere 62: 1411–1420.

Liu, P.W.G., T.C. Chang, L.M. Whang, C.H. Kao, P.T. Pan and S.S. Cheng. 2011. Bioremediation of petroleum hydrocarbon contaminated soil: Effects of strategies and microbial community shift. Int. Biodeter. Biodegr. 65: 1119–1127.

Lladó, S., S. Covino, A.M. Solanas, M. Viñas, M. Petruccioli and A. D'annibale. 2013. Comparative assessment of bioremediation approaches to highly recalcitrant PAH degradation in a real industrial polluted soil. J. Hazard. Mater. 248-249: 407–414.

Magalhães, S.M.C., R.M.F. Jorge and P.M.L. Castro. 2009. Investigations into the application of a combination of bioventing and biotrickling filter technologies for soil decontamination processes—A transition regime between bioventing and soil vapour extraction. J. Hazard. Mater. 170: 711–715.

Mohsenzadeh, F., S. Nasseri, A. Mesdaghinia, R. Nabizadeh, D. Zafari, G. Khodakaramian and A. Chehregani. 2010. Phytoremediation of petroleum-polluted soils: Application of Polygonum aviculare and its root-associated (penetrated) fungal strains for bioremediation of petroleum-polluted soils. Ecotox. Environ. Safe. 73: 613–619.

Mrozik, A. and Z. Piotrowska-Seget. 2010. Bioaugmentation as a strategy for cleaning up of soils contaminated with aromatic compounds. Microbiol. Res. 165: 363–375.

Österreicher-Cunha, P., E.A. Vargas, Jr., J.R.D. Guimarães, T.M.P. de Campos, C.M.F. Nunes, A. Costa, F.S. Antunes, M.I.P. da Silva and D.M. Mano. 2004. Evaluation of bioventing on a gasoline–ethanol contaminated undisturbed residual soil. J. Hazard. Mater. 110: 63–76.

Peng, S., Q. Zhou, Z. Cai and Z. Zhang. 2009. Phytoremediation of petroleum contaminated soils by Mirabilis Jalapa Lin a greenhouse plot experiment. J. Hazard. Mater. 168: 1490–1496.

Steliga, T., P. Jakubowicz and P. Kapusta. 2012. Changes in toxicity during *in situ* bioremediation of weathered drill wastes contaminated with petroleum hydrocarbons. Biores. Technol. 125: 1–10.

Suja, F., F. Rahim, M.R. Taha, N. Hambali, M.R. Razali, A. Khalid and A. Hamzah. 2014. Effects of local microbial bioaugmentation and biostimulation on the bioremediation of total petroleum hydrocarbons (TPH) in crude oil contaminated soil based on laboratory and field observations. Int. Biodeter. Biodegr. 90: 115–122.

Winquist, E., K. Björklöf, E. Schultz, M. Räsänen, K. Salonen, F. Anasonye, T. Cajthaml, K.T. Steffen, K.S. Jørgensen and M. Tuomela. 2014. Bioremediation of PAH-contaminated soil with fungi e From laboratory to field scale. Int. Biodeter. Biodegr. 86: 238–247.

# CHAPTER 10

# Phytoremediation of Salt Affected Soils

*João M. Jesus,[1] Anthony S. Danko,[1,*] António Fiúza[1]*
and *Maria-Teresa Borges[2,3]*

## ABSTRACT

Salt affected soils occupy up to 10 billion hectare of soil and decrease agricultural productivity, threatening food security worldwide. Leaching and chemical amendments, the most used remediation techniques, can have limited applications and heavy costs. Phytoremediation is a viable, less expensive alternative, with opportunities for added value. However, its application requires a deeper knowledge of soil and plant interactions, as well as of existing resources. This chapter will analyze the essential concepts of soil salinity and sodicity, as well as irrigation water sodicity hazard, and how to manage different combinations of these parameters. Furthermore, the most relevant soil parameters for plant utilization are identified and the performance of phytoremediation applied to salt affected soils under field conditions is evaluated, as well as possible limiting factors. A practical view of plant salt tolerance and uptake is shown which is correlated with performance and applicability of this technique.

[1] Centre for Natural Resources and the Environment (CERENA), Department of Mining Engineering, University of Porto - Faculty of Engineering (FEUP), Rua Dr. Roberto Frias s/n, 4200-465, Porto, Portugal.
[2] Biology Department, Science Faculty, Porto University (FCUP), Rua Campo Alegre s/n, 4169-007 Porto, Portugal.
[3] CIIMAR, University of Porto, Rua dos Bragas 289, 4050-123 Porto, Portugal.
* Corresponding author: asdanko@fe.up.pt

## Extent and Worldwide Distribution of Soil Salinization

Soil salinization can be described as the excess of salts and/or of sodium ions, either in the soil solution or in its cation exchange sites (Qadir et al. 2000). The main causes for this accumulation of salts in the soil profile derives from natural events (geological deposition, saline groundwater) or is anthropogenically formed or is enhanced by degradation processes. These include loss of vegetative cover, poor irrigation schemes, saline wastewater, saline intrusion due to aquifer overexploitation, etc. A saline soil is characterized by an electrical conductivity of the saturated soil paste (ECe) above 4 dS m$^{-1}$ and a sodium adsorption ratio (SAR) below 13. A sodic soil, on the other hand, is characterized by an ECe under 4 dS m$^{-1}$ and a SAR above 13. A saline-sodic soil is a combination of the previous ones with an ECe above 4 dS m$^{-1}$ as well as a SAR above 13 (US Soil Salinity Laboratory Staff 1954; Qadir et al. 2000).

Soil salinization affects several continents (Table 1) and over 100 different countries, having a conservatively estimated impact of 12 billion dollars per year worldwide (Pitman and Läuchli 2004). While it is estimated that food production needs to be increased by 50% by the year 2050 (Rengasamy 2006), at the same time the area available for agriculture expansion is limited due to salinization and other problems. Recovering areas not suited for agriculture such as salt affected soils, may be essential for sustainable development. Asia is the continent (particularly Central Asia) with the highest area occupied by saline and sodic soils (Table 1), a result of a combination of intense irrigation schemes and arid and semi-arid climate that is dominant in the region.

**Table 1.** Worldwide distribution of salt affected soils—saline and sodic soil distribution by (sub) continent (based on Szabolcs 1989; Abrol et al. 1988).

| Continent | Saline soils (Mha) | Sodic soils (Mha) |
|---|---|---|
| North America | 6.2 | 9.6 |
| South America | 69.4 | 59.6 |
| Africa | 53.5 | 27.0 |
| Australasia | 17.4 | 340.0 |
| North and Central Asia | 91.6 | 120.1 |
| South Asia | 83.3 | 1.8 |
| Europe | 7.8 | 22.9 |
| Total | 329.2 | 581.0 |

The total area of salt affected soil worldwide is close to 910 Mha, which is equivalent to 60% of all arable area or under permanent crops and with an estimated annual increase of up to 16% (Metternicht and Zinck 2008). Close to 76.3 Mha of these soils are classified as resulting from human induced salinization, of which 55% are of moderate to extreme salinity. This anthropogenic induced soil degradation problem is more severe than pollution, acidification, compaction, and water logging (Metternicht and Zinck 2008).

This book chapter shall focus on the remediation of saline soils and will explore the basic and advanced technical knowledge necessary for appropriate remediation efforts. A special emphasis will be given to salt phytoremediation, debating some examples of its application in the field.

## Basic Salinization Monitoring Parameters and Techniques Used

In salt affected soils salinity is, obviously, a paramount characteristic. However, as shall be discussed below, there are diverse methods that can be applied to assess salinity, and their significance and applicability *in situ* varies considerably. It is, therefore, important to know under which circumstances each method is the most suited. Furthermore, sodicity is equally important in the assessment and remediation of salt affected soils. In addition, there is the need to adapt different techniques to the precision required in each case.

### Electrical Conductivity

In any study of salt affected soils, the most important parameter to be determined is salinity, which is usually monitored by electrical conductivity (EC). There is an immense variety of methods that can be used to determine EC, with different degrees of applicability and range, depending on the objective of the studies that are being conducted. The methods applied for electric conductivity determination can be very diverse. The most advanced techniques such as bulk soil electrical conductivity (ECa) are mostly used in *in situ* measurements for mapping (Rhoades et al. 1999). Although this is one of the most accurate measuring methods, it is more costly and therefore it is less used.

The electric conductivity of saturated soil paste (ECe) however, is widely used as a standard method. A saturated paste is prepared by adding distilled water to the soil until a uniform saturated soil paste is achieved. At this point, since it is considered sufficiently reproducible, the water is extracted and the EC and temperature are measured (US Soil Salinity Laboratory Staff 1954; Qadir 2000). A simpler method, but more prone to errors, is soil to water dilutions in varying ratios of 1:1 up to 1:5. It is mostly valid when relative changes in EC, and not its absolute value, are of more concern. Some approximations have been proposed using texture, since ECe depends on this parameter. This is done by multiplying EC by a factor of 12.5 for sandy soils, by 10 for sandy loam, by 8 for loam or clay loam, by 7 for light clay or by 6 for heavy clay soils (US Soil Salinity Laboratory Staff 1954).

The reasons for preferring these methods (ECe and soil to water dilutions) over the more advanced ones are the need of less expensive equipment that provides more accurate (although much less representative) data, and the fact that collected samples can be used for the assessment of other relevant parameters such as SAR and available nutrients.

Converting electric conductivity to TDS (total dissolved salts), or vice-versa, is sometimes required but can be difficult to do as electric conductivity is dependent on sample cation and anion composition. Several simple rules for this conversion exist for soil applications, the most common of which is EC = TDS/0.640, where TDS is in g L$^{-1}$ or g kg$^{-1}$ and EC in dS m$^{-1}$ (US Soil Salinity Laboratory Staff 1954; Qadir et al. 2000). Although the calculations are simplified, this conversion is widely used and accepted as a rule of thumb.

Common errors or misinterpretations of the electrical conductivity of salt affected soils can be derived from poor sampling techniques (not statistically representative) or misapplication of measuring techniques. For instance, the EC 1:5 methodology requires filtration or centrifugation in specific soil types. Direct supernatant EC measurement can result in a severely underestimated salinity value or, sometimes, just an inaccurate result representation (EC 1:1 is sometimes equated to ECe, for example).

### SAR and ESP Levels

The other most important factor to be monitored in salt affected soils is sodicity. Sodicity represents the level of adsorbed sodium in a soil in relation with other cations and can be assessed by determining SAR and/or ESP (Exchangeable Sodium Percentage). SAR establishes the relationship between soluble sodium and divalent cation concentration (magnesium and calcium cations) (US Soil Salinity Laboratory Staff 1954):

$$SAR = \frac{Na^+}{\sqrt{\dfrac{Ca^{2+} + Mg^{2+}}{2}}} \quad \text{or} \quad SAR = \frac{Na^+}{\sqrt{Ca^{2+} + Mg^{2+}}}$$

where Na$^+$, Ca$^{2+}$ and Mg$^{2+}$ refer to the concentrations of respective ions, in meq L$^{-1}$ in the first equation, and mmol L$^{-1}$ in the second one. This parameter can be evaluated experimentally by the determination of soluble Na$^+$, Ca$^{2+}$, and Mg$^{2+}$ concentrations in saturated paste extracts. After analyzing the existing scientific literature, it can be said that the most commonly used methods for the determination of these cations are potentiometry using ion selective electrodes, atomic absorption spectrometry, flame photometry, and ion chromatography. Additionally, EDTA titration can be used for calcium and magnesium.

ESP is also used to express sodicity. Since it represents exchangeable sodium, ESP is a more accurate assessment method of sodicity when compared with SAR, and can be defined as the coefficient between exchangeable sodium and total Cation Exchange Capacity (CEC) of the soil (Wong et al. 2009):

$$ESP = \frac{Na^+}{CEC}; ESP = \frac{Na^+}{Ca^{2+} + Mg^{2+} + K^+ + Na^+}$$

It can be determined experimentally but it is a long and laborious process and prone to errors; therefore, ESP is normally estimated based on SAR values using the following equation (US Soil Salinity Laboratory Staff 1954):

$$ESP = [100\,(-0.0126 + 0.01475\,SAR)]/[1 + (-0.0126 + 0.01475\,SAR)]$$

### Irrigation Water Sodicity Hazard

Evaluation of sodicity hazard of irrigation water by measuring SAR alone can prove to be insufficient. This happens because the calculation of SAR does not take into account ion pair formation due to the presence of carbonate and bicarbonate, which reduces the activity of calcium and magnesium cations (Qadir and Schubert 2002).

To correct for this factor, $SAR_{adj}$ may be used (Suarez 1981) in which calcium levels in the SAR calculation, in mmol $L^{-1}$, are substituted by an equivalent concentration. This concentration is calculated through the following equation:

$$Ca^{2+}_{eq} = X * (Pco_2)^{\frac{1}{3}}$$

in which X is approximately determined using the look-up table developed by Suarez (1981) and the calculated ratio of $HCO_3^-/Ca^{2+}$ and ionic strength.

Alternatively, the Residual Sodium Carbonate (RSC) can also be used, being calculated through the following equation (Carrow and Duncan 2011):

$$RSC = (CO_3^{2-} + HCO_3^-) - (Ca^{2+} + Mg^{2+})$$

where all ion concentrations are expressed in meq $L^{-1}$. If RSC is negative, the irrigation water can be beneficial to control the soil sodicity, if low in sodium ions. If it is positive, however, it may precipitate existing calcium in the soil, increasing SAR levels (Carrow and Duncan 2011). RSC, however, fails to account for existing sodium concentration and therefore should be assessed in conjunction with non-adjusted SAR data. Information on the RSC levels and of their influence on irrigation water is applied to establish limits for groundwater recovery, as well as for analyzing the quality of the water to be applied for soil remediation.

## Water and Soil: Interactions and Management Options Under Saline Conditions

When considering any type of salt affected soil remediation (either leaching, chemical or organic amendment, electrokinetics or phytoremediation) available water quality always has to be taken into consideration. Technical and/or economic feasibility of remediation is dependent on both soil and irrigation water quality (Fig. 1) (Qadir and Oster 2002). Irrigation water can be classified as fresh, brackish, or highly saline water but the values used to define these terms vary in the literature and are summarized in Table 2.

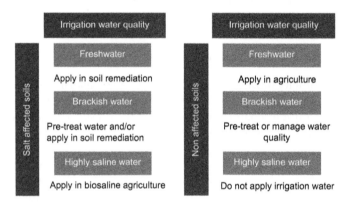

**Fig. 1.** Irrigation water management options according to water quality as well as soil quality.

**Table 2. Irrigation water type**—Sodicity hazard dependence on EC, $SAR_{adj}$ and RSC values. (US Soil Salinity Laboratory Staff 1954; Oster and Schroer 1979.)

| Water type | Risk | EC (dS m$^{-1}$) | $SAR_{adj}$ | RSC (meq L$^{-1}$) | *$SAR_{adj}$/EC (dS m$^{-1}$) |
|---|---|---|---|---|---|
| Freshwater | No risk | < 0.5 | < 10 | < 0 | 0/> 0.7 |
| | | | | | 6/> 1.9 |
| | | | | | 20/> 5 |
| Brackish water | Medium Risk | 0.5–3.0 | 10–18 | 0–1.25 | 0/0.2–0.7 |
| | | | | | 6/0.5–1.9 |
| | | | | | 20/2.9–5 |
| Highly saline water | High Risk | > 3.0 | > 18 | > 1.25 | 0/< 0.2 |
| | | | | | 6/< 0.5 |
| | | | | | 20/< 2.9 |

\* $SAR_{adj}$/EC—$SAR_{adj}$ value associated with a corresponding EC value.

In soils not affected by salt, irrigation water should be of the highest quality available to enable agriculture practice of traditional crops. However, with some adaptations, brackish water can and is used to irrigate more tolerant crops in less susceptible soils, such as soils with low CEC (Qadir et al. 2001). There is a great variety of techniques available to diminish the potential impact of brackish water irrigation, that range from sub irrigation to blending with other water sources, or seasonal use only as is, for example, the use of brackish water in the summer when freshwater resources are scarcer (Murtaza et al. 2006). However, the efficiency of these techniques is limited and highly saline irrigation water should ideally never be applied in agricultural soils.

In case soils are already salt affected, the quality of irrigation water affects the efficiency or even the feasibility of the remedial process. Ideally, regardless of the remedial technique, the irrigation water should be of high quality. However, brackish water can also be used if applied at a sufficient rate to leach salts and therefore it has an overall positive effect on the soil salt balance.

In the case of highly saline water, remediation techniques that are technically capable to treat salt affected soils in those conditions are more limited and potentially cost prohibitive. In such cases, growth of extreme halophytes of added value (such as medicinal, ornamental, fodder for animals, etc. (Cassaniti et al. 2012)) could stabilize the soil (although with little to no salt remediation), creating value in soils that otherwise would not be used. Even in these cases, sodicity control is necessary to maintain adequate soil drainage, as well as to avoid leaching salts to productive aquifers, if existing in the region.

Another approach to the use of brackish water is to pre-treat it before use. This opens up the possibility of reusing saline wastewater or combining salt remediation with other required treatments. Such simultaneous treatment could be achieved by Constructed Wetlands (CWs) (Shelef et al. 2012). CWs could also be used to remove salts from more diffuse, non-point sources, or it can even be applied for groundwater remediation. Although there are several studies on wastewater treatment by constructed wetlands under saline conditions (see review from Buhmann and Papenbrock 2012), few studies focus on salt removal rather than salt tolerance. Certain studies, though, have reported salt co-treatment with other pollutants (Lymbery et al. 2006; Jesus et al. 2013), whereas other types of constructed wetlands were designed for the sole purpose of salt removal, with auspicious effects (Shelef et al. 2012).

Figure 1 can be used as a rule of thumb to help decide when remediation (in particular phytoremediation) is a viable, cost effective option to be considered, either for salt affected soil or water, or when management and control of the salinization problem is a more realistic and sustainable approach.

## Phytoremediation of Salt Affected Soils

### Site Characterization Needs

Prior to any remediation project, or even of a treatability test in the specific case of a salt affected soil, a thorough and detailed site characterization is necessary for adequate management. Besides the salinity and sodicity indexes already discussed, a set of relevant parameters need to be taken into account, which are summarized in Table 3.

**Table 3. Soil and climatic characteristics**—relevant parameters for site assessment and their respective impact on soil characteristics influencing salt affected soil remediation (US Soil Salinity Laboratory Staff 1954; Qadir et al. 2001; AENV 2001; Gawel 2006).

| Parameter | Impact |
|---|---|
| Soil Texture | Texture determines water retention and flow, as well as salt adsorption. |
| Soil Clay type | Different types have different salt adsorption rates. |
| Soil $CaCO_3$ | Native calcium sources should be taken into account as possible aids. |
| Soil Gypsum | Natural gypsum deposits may enhance remediation where available. |
| Soil CEC | Regulates the maximum amount of adsorbed cations. |
| Evaporation | Determines rate of capillary rise and salt accumulation. |
| Precipitation | Limits water availability for leaching. |
| Air Temperature | Pertinent for plant growth and salt dissolution in applied water. |
| Water table | If too high, leachate collection is required. |
| Irrigation water quality | Water quality should be compatible with remediation goals. |
| Initial soil pH | Important for plant growth and salt bioavailability. |
| Soil Slope | If high, leaching may create saline seepage. |

Native sources of calcium ions of different solubility ($CaCO_3$ and gypsum) may provide opportunities for increased remediation efficiency if pH is reduced during remediation (which would increase their dissolution). These mineral deposits can be mobilized and homogenously distributed by tillage techniques like deep ploughing or sub soiling (Qadir et al. 2001), with the added benefits of improving soil structure.

Climatic conditions are also extremely relevant, especially in arid and semi-arid regions where precipitation is not only limited but is vastly surpassed by evaporation, leading to increased salt capillary rise (AENV 2001).

If salt leaching is to be used, further restrictions may apply. If the water table is high, leaching would not only transport salts to the groundwater but would also, through capillary rise, rapidly recontaminate the soil. Furthermore, if the terrain has a significant slope, saline seepage may occur in valley regions, transferring, rather than treating, the salinity problem (Gawel 2006).

Although outside the scope of this chapter, appropriate sampling campaigns of statistical relevance using adequate methods are vital for data interpretation.

### Plant Salt Tolerance and Uptake—A Mechanistic Approach

There exists a large variety of classifications for plant salt tolerance and detailed descriptions of the biochemical and metabolic pathways enabling this tolerance (Huchzermeyer and Flowers 2013). However, in this chapter, it was chosen to give a focus on valuable practical information on plants adequate for salt remediation.

The first point is to clarify plant classification regarding the presence of salt. One of the most known and basic classification divides plants into three categories: glycophytes, halotolerant, and halophytes. These three categories are simple to define: glycophytes are the most sensitive and halophytes the most tolerant to salinity. Halophytes, however, can be further classified into obligate, semi-obligate, or facultative halophytes, ordered in decreasing dependence on salinity for stimulated growth (Huchzermeyer and Flowers 2013).

These classifications, however, do not provide any information on salt uptake. As a result, Yensen and Biel (2006) suggested a new classification that divides plant response to salt into accumulators, excluders, and conductors. Excluder type plants, regardless of their salt tolerance, actively prevent salt from entering through the roots. Accumulator plants, on the other hand, store salts in their vacuoles to prevent toxicity and maintain adequate osmotic pressure. Finally, conductor plants uptake salts but also excrete them through salt bladders or glands in the leaves or shoots.

In the case of accumulators, not only is the whole plant salt uptake relevant, but also where salt accumulates. The translocation factor, commonly used for heavy metals, is also applied for salts, indicating the ratio between the salts in above ground and below ground biomass (Guittonny-Philippe et al. 2014). Although translocation to shoots is desirable, roots with accumulated salts can also be partly removed without affecting plant survival. This scenario is particularly true in CWs where plant roots and shoots can be easily harvested (Guittonny-Philippe et al. 2014).

Total biomass productivity is also extremely relevant for salt balance in salt remediation situations, as seen for metal phytoremediation. For instance, (Goulet et al. 2005) found that *Lemna minor* had an aluminum concentration close to five times that of *Typha latifolia*, although the latter was responsible for 99% of the aluminum removed in the conducted mesocosm tests. The same could be applied in salt remediation scenarios.

Halophytic crop production may also be an added value opportunity resulting from the salt remedial process. In such cases, the choice of plants and management options to increase yield depends on the treatment goals. These crops can have a multitude of uses but are more traditionally used as fodder for animals. The hemicellulose and cellulose can also be converted to ethanol, biogas, and hydrogen (Kaparaju et al. 2009).

However, plants used in phytoremediation may have undesirable constituents as a result of their contribution to the remediation process. For instance, plants with a high salt uptake may inhibit fermentation due to excess sodium, chloride, or potassium or could have negative health effects on animals if used as fodder (Walton et al. 2010). Consequently, if such uses are planned for the vegetation employed in the remedial process, a maximum salt concentration within plant tissue should be established, forcing more frequent harvests, or the use of conductor plants rather than accumulators.

### *Treatment Efficiency*

Laboratory and treatability studies on the efficiency of phytoremediation of salt affected soils can provide important and relevant information for future applications. Nonetheless, field studies can more closely resemble the actual remedial conditions and provide robust information on the effects of uncontrollable parameters such as climate.

Due to their large scale and time span, as well as several other restrictions of logistical nature, field studies are far scarcer than lab scale tests. Table 4 summarizes the initial conditions of four different field case studies that will be analyzed to illustrate the treatment performance in the field and the constraints that may apply.

In the first situation in Table 4 (Gansu Province, China), field experiments with *Medicago sativa*, harvested twice a year, revealed a yearly EC reduction of 31%, 27%, 10%, 3%, and 5% during the five years of this study. In the same period, plant yield was approximately 10, 15, 20, 23, and 28 ton per ha, respectively. Both datasets indicate some initial salt sensitivity in plants, possibly due to a higher initial salt accumulation (indicated by higher EC reduction). Lower EC values led to an increased yield, while the reverse was not observed, possibly indicating that, at some time-point, soil EC value was already too low for visible changes in this parameter, namely because by the second year the soil could already be considered non saline (< 4 dS m$^{-1}$). An increase in total organic carbon, available P and total N represent a positive side-effect of the remediation process as well, since they are contributing factors for increasing plant yields. The produced biomass was used exclusively as forage for farm animals as an added value by-product.

In the work of Keiffer and Ungar (2002), however, site conditions were significantly different. The site was a former pasture ground, severely affected by an accidental, yearlong, brine collection tank leakage of undocumented composition. As in the previous study, though, the selected plant species were sown instead of transplanted, to simulate realistic implementation scenarios. Of the five tested plants, two (*Spergularia marina* and *Atriplex prostrata*) had a significantly greater germination, survival, and yield rate when sown in spring,

Table 4. Site conditions of field tests—Soil characteristics of four sites tested for salt affected soil phytoremediation.

| Site and Source | (Cao et al. 2012) | (Keiffer and Ungar 2002) | (Ghaly 2002) | (Akhter et al. 2004) |
|---|---|---|---|---|
| Location | Gansu Province, China | Ohio, USA | Northern Egypt | Lahore, Pakistan |
| Air Temperature (°C) | 5.9 | 7.5 | - | - |
| Precipitation (mm) | 261 | 117 | - | 497 |
| Soil Texture | Loam | Clay loam | > 56% clay | Sandy clay loam |
| Soil CaCO$_3$(%) | 12.8 | - | 1.4 | 0.6 |
| Soil pH | 8.1 | - | 8.1 | 10.4 |
| Soil EC (dS m$^{-1}$) | 10 | 25 | 25.4 | 22.0 |
| SAR | - | 40 meq L$^{-1}$ Na$^+$ | 24.5 | 184 |
| Time (years) | 5 | 2 | 2 | 5 |
| Plot Area (m$^2$) | 200 | 1.5 | 7 | 900 |

while one of the plants (*Hordeum jubatum*) performed better when sown in autumn. The remaining two, *Salicornia europaea* and *Suaeda calceoliformis*, had no different results. Maximum removal (based on the direct effect of plant uptake only) of 17% Na$^+$ was obtained with *A. prostrata* when sown in the spring, with a homogenous distribution of salt uptake throughout the roots, stems, and leafs. These results were obtained without irrigation, and so leaching was limited to rainfall. This field test further exemplifies the need to know the plant life cycle to improve phytoremediation performance and pinpoints one of the limiting factors for its implementation: the lack of commercially available halophyte seeds.

In the field test referred by Ghaly (2002), a moderately saline-sodic soil was treated by different sets of treatments: leaching, gypsum, two plant species (*Phragmites communis* and *Panicum repens*) with or without gypsum addition. In the first year, EC was reduced by up to 80% and SAR level by 33% with *P. repens*, resulting in a soil remediation close to non-saline soils. Tested plants either outperformed or had similar removal efficiency than chemical and physical treatments in the same time span. However, several treatment combinations were tested, and those that combined gypsum with plants had increased salt removal efficiency and plant uptake.

In the last case study analyzed (Akhter et al. 2004), the soil was extremely sodic (SAR = 184) and plant irrigation was carried out by a low quality groundwater (with SAR$_{adj}$ = 19.3; EC = 0.14 dS m$^{-1}$ and RSC = 9.7 meq L$^{-1}$). Furthermore, the initial soil permeability was so low that it prevented the existence of a leaching control plot. After five years of testing, SAR was reduced by 84% and EC by 89% with the growth of *Leptochloa fusca*. Adequate hydraulic conductivity and soil structural stability were achieved in three years and maintained thorough the remaining two years of the study.

Finally, it must be stressed that two of these case studies were performed in regions of monsoons (Akhter et al. 2004; Cao et al. 2012), in which up to 80% of all precipitation occurs in just two to three months of the year and therefore plants may have experienced, and tolerated, flood conditions, additionally testifying the robustness and adaptability of this remedial method. There is a lack, however, of comparable studies in more temperate areas and further studies should be conducted to evaluate performance under such climatic conditions.

### Applicability Limitations and Constraints

Not only is it important to currently assess whether or not remediation is a cost effective possibility, but also what are the limits of phytoremediation applicability.

The following parameters can be considered as limiting factors for the application of phytoremediation:

> **Time**—The potential urgency for remediation for whatever reason precludes the use of phytoremediation, as well as of most other *in situ* remediation options. In extremely severe, yet localized cases of soil salinization, *ex situ* techniques may be required for a quick recovery (AENV 2001).

> **Treatment depth**—Treatment depth is limited to the root zone although, in some cases, the effective depth remediated by phytoremediation can be higher than in other treatments, namely with gypsum (Qadir and Oster 2002).

> **Initial SAR/Sodium level**—No matter how resistant they may be, all plants eventually will suffer the toxic effects of excessive sodium. Toxicity levels may vary but a threshold concentration, especially if combined with other stress conditions, is always present in any plant species. If that high level of sodium is accompanied by high SAR, it might render phytoremediation unmanageable (Qadir and Oster 2002).

> **Soil structure**—Excessive adsorbed sodium is known to cause soil dispersion and impermeabilization. At extreme levels, and in soils with high clay content, root penetration may be impossible, as well as leaching. Remediation, in such cases, must be preceded by expensive physical and vigorous mixing of the soil to break impermeable layers, if possible. There are reported cases of geological deposition from ancient seas in which soil could only be excavated with dynamite (Singh 2003).

**Removal mechanism**—The main removal mechanism by which specific phytoremediation schemes may be based upon have different ranges of applicability. If phytoremediation is planned as a technique to decrease pH and mobilizes native calcium sources (Qadir and Oster 2002) to improve soil structure and salt leaching, sites in which there are little native calcium ions in the soils and with low water availability for leaching, may prevent efficient salt removal rates. On the other hand, phytoremediation schemes based on plant salt uptake (Rabhi et al. 2010) do not suffer from these limitations, although their remediation time span is larger. These schemes can be more economical as they may avoid the need for leachate collection and disposal.

**Economic**—Several economic parameters, although not affecting technical feasibility, may influence the choice of whether or not to remediate and which particular technique is the most cost-effective in specific cases. For instance, when plans for use of phytoremediation vegetative biomass for animal fodder are considered, an analysis of the regional market for this specific product may potentially make up for large remediation costs and in some reported cases, provide a positive economic balance (Qadir and Oster 2002). Variability in chemical amendment prices may also impact remediation choices.

In summary, not unlike other remediation techniques, phytoremediation for salt affected soils has a finite range of applicability (although its limits are still unclear) and should only be considered when case specific parameters are properly taken into account. In certain situations, combined use with other techniques may prove to be necessary and/or have synergetic effects on treatment efficiency and economic feasibility.

## Conclusion

Phytoremediation, and in particular phytoextraction, is yet a poorly explored option in remediation sites, but is having an ever growing acceptance among practitioners and stakeholders. In the case of salt affected soils, phytoremediation has been extensively and successfully applied in many Middle Eastern countries. However, the soil salinization problem, due to poor irrigation schemes and climate change, is rapidly expanding to other parts of the world and the use of the full, unexplored potential of phytoremediation of salt affected soils will be required to face new challenges and circumstances, such as remediation of non-calcareous soils and under non-leaching conditions.

For that purpose, an accurate and representative performance analysis, as well as the knowledge of the limitations of the technique, is of extreme importance.

Therefore, this chapter delivered information, as detailed as possible, on the performance of field tests of phytoremediation with recommendations for case specific decision making tools and practical examples.

## Acknowledgements

The authors would like to acknowledge the Portuguese Science and Technology Foundation (FCT) for the PhD grant (FCT—DFRH—SFRH/BD/84750/2012) and the Ciência 2008 program.

## References

Abrol, I.P., J.S.P. Yadav and F.I. Massoud. 1988. Salt affected soils and their management. FAO Soils Bull. 39, FAO, Rome, 131.

Akhter, J., R. Murray, K. Mahmood, K.A. Malik and S. Ahmed. 2004. Improvement of degraded physical properties of a saline-sodic soil by reclamation with kallar grass (*Leptochloa fusca*). Plant Soil 258: 207–216.

[AENV] Alberta Environment 2001. Salt contamination assessment and remediation guidelines. Environmental Sciences Division.

Buhmann, A. and J. Papenbrock. 2012. Biofiltering of aquaculture effluents by halophytic plants: Basic principles, current uses and future perspectives. Environ. Exp. Bot. 92: 122–133.

Cao, J., X. Li, X. Kong, R. Zed and L. Dong. 2012. Using alfalfa (*Medicago sativa*) to ameliorate salt-affected soils in Yingda irrigation district in Northwest China. Acta Ecol. Sin. 32: 68–73.

Carrow, R.N. and R.R. Duncan. 2011. Best Management Practices for Saline and Sodic Turfgrass Soils: Assessment and Reclamation. Taylor & Francis, London, UK.

Cassaniti, C., D. Romano, M.E.C.M. Hop and T.J. Flowers. 2012. Growing floricultural crops with brackish water. Environ. Exp. Bot. 92: 165–175.

Gawel, L.J. 2006. A Guide for Remediation of Salt/Hydrocarbon Impacted Soil. North Dakota Industrial Commission, Department of Mineral Resources, Bismarck, ND.

Ghaly, F.M. 2002. Role of natural vegetation in improving salt affected soil in northern Egypt. Soil Tillage Res. 64: 173–178.

Goulet, R.R., J.D. Lalonde, C. Munger, S. Dupuis, G. Dumont-Frenette, S. Prémont and P.G.C. Campbell. 2005. Phytoremediation of effluents from aluminum smelters: A study of Al retention in mesocosms containing aquatic plants. Water Res. 39: 2291–2300.

Guittonny-Philippe, A., V. Masotti, P. Höhener, J.-L. Boudenne, J. Viglione and I. Laffont-Schwob. 2014. Constructed wetlands to reduce metal pollution from industrial catchments in aquatic Mediterranean ecosystems: A review to overcome obstacles and suggest potential solutions. Environ. Int. 64: 1–16.

Huchzermeyer, B. and T. Flowers. 2013. Putting halophytes to work—genetics, biochemistry and physiology. Funct. Plant Biol. 40: v–viii.

Jesus, J.M., C.S.C. Calheiros, P.M.L. Castro and M.T. Borges. 2013. Feasibility of *Typha Latifolia* for high salinity effluent treatment in constructed wetlands for integration in resource management systems. Int. J. Phytorem. 16: 334–346.

Kaparaju, P., M. Serrano, A.B. Thomsen, P. Kongjan and I. Angelidaki. 2009. Bioethanol, biohydrogen and biogas production from wheat straw in a biorefinery concept. Bioresour. Technol. 100: 2562–2568.

Keiffer, C.H. and I.A. Ungar. 2002. Germination and establishment of halophytes on brine-affected soils. J. Appl. Ecol. 39: 402–415.

Lymbery, A.J., R.G. Doupé, T. Bennett and M.R. Starcevich. 2006. Efficacy of a subsurface-flow wetland using the estuarine sedge Juncus kraussii to treat effluent from inland saline aquaculture. Aquacult. Eng. 34: 1–7.

Mateo-Sagasta, J. and J. Burke. 2012. Agriculture and water quality interactions: A global overview. SOLAW Background Thematic Report—TR08.

Metternicht, G. and A. Zinck. 2008. Remote Sensing of Soil Salinization: Impact on Land Management. Taylor & Francis, London, UK.

Murtaza, G., A. Ghafoor and M. Qadir. 2006. Irrigation and soil management strategies for using saline-sodic water in a cotton–wheat rotation. Agric. Water Manage. 81: 98–114.

Oster, J.D. and F.W. Schroer. 1979. Infiltration as influenced by irrigation water quality. Soil Sci. Soc. Am. J. 43: 444–447.

Pitman, M. and A. Läuchli. 2004. Global impact of salinity and agricultural ecosystems. pp. 3–20. *In*: A. Läuchli and U. Lüttge (eds.). Salinity: Environment—Plants—Molecules. Springer, The Netherlands.

Qadir, M. and J. Oster. 2002. Vegetative bioremediation of calcareous sodic soils: History, mechanisms, and evaluation. Irrigation Sci. 21: 91–101.

Qadir, M. and S. Schubert. 2002. Degradation processes and nutrient constraints in sodic soils. Land Degrad. Dev. 13: 275–294.

Qadir, M., A. Ghafoor and G. Murtaza. 2000. Amelioration strategies for saline soils: A review. Land Degrad. Dev. 11: 501–521.

Qadir, M., S. Schubert, A. Ghafoor and G. Murtaza. 2001. Amelioration strategies for sodic soils: A review. Land Degrad. Dev. 12: 357–386.

Rabhi, M., S. Ferchichi, J. Jouini, M.H. Hamrouni, H.-W. Koyro, A. Ranieri, C. Abdelly and A. Smaoui. 2010. Phytodesalination of a salt-affected soil with the halophyte *Sesuvium portulacastrum* L. to arrange in advance the requirements for the successful growth of a glycophytic crop. Bioresour. Technol. 101: 6822–6828.

Rengasamy, P. 2006. World salinization with emphasis on Australia. J. Exp. Bot. 57: 1017–1023.

Rhoades, J.D., F. Chanduvi and S. Lesch. 1999. Soil salinity assessment: Methods and interpretation of electrical conductivity measurements. FAO Irrigation and Drainage Paper #57. Food and Agriculture Organization of the United Nations, Rome, Italy 1–150.

Shelef, O., A. Gross and S. Rachmilevitch. 2012. The use of *Bassia indica* for salt phytoremediation in constructed wetlands. Water Res. 46: 3967–3976.

Singh, D.P. 2003. Stress Physiology. New Age International (P) Ltd., New Delhi, India.

Suarez, D.L. 1981. Relation between pHc and sodium adsorption ratio (SAR) and an alternative method of estimating SAR of soil or drainage waters. Soil Sci. Soc. Am. J. 45: 469–475.

Szabolcs, I. 1989. Salt-Affected Soils. CRC Press, Boca Raton, USA.

US Soil Salinity Laboratory Staff 1954. Diagnosis and Improvement of Saline and Alkali Soil. Agric. Handbook. Dept. of Agric. Washington, DC, USA.

Walton, S., A. van Heiningen and P. van Walsum. 2010. Inhibition effects on fermentation of hardwood extracted hemicelluloses by acetic acid and sodium. Bioresour. Technol. 101: 1935–1940.

Wong, V.N.L., R.C. Dalal and R.S.B. Greene. 2009. Carbon dynamics of sodic and saline soils following gypsum and organic material additions: A laboratory incubation. Appl. Soil Ecol. 41: 29–40.

Yensen, N.P. and K.Y. Biel. 2006. Soil remediation via salt-conduction and the hypotheses of halosynthesis and photoprotection. pp. 313–344. *In*: M.A. Khan and D.J. Weber (eds.). Ecophysiology of High Salinity Tolerant Plants. Springer, The Netherlands.

# Index

3-phenoxybenzaldehyde 62, 64–68
3-phenoxybenzoic acid 62, 64, 65

**A**

activated persulfate 75, 78, 79, 85
Aquifer Thermal Energy Storage 121, 123, 128

**B**

bioaugmentation 24, 26, 27, 34, 35, 139, 142, 143, 145, 146
biological remediation 139, 140, 145, 146
bioremediation vii, 59, 63, 64, 69, 71, 141–146
Biostimulants 21, 24, 28–35
biostimulation 24, 27, 28, 35, 139, 141–143, 146
bioventing 143

**C**

carboxylesterase 70, 71
Chemical Oxidation vii
Chlorinated ethenes 21–25, 27–29, 36
Cometabolism 22, 23, 27
Contaminated groundwater 121, 122, 124–127, 129, 136, 137
contaminated sites 95, 97, 98, 102–105
contamination v, vii, 95, 96, 98–100, 104, 105
contamination plumes 122
Co-solvents 11, 13
Cyclodextrins 11, 13–15

**D**

*Dehalococcoides* 24, 34
Dosing 21, 30, 31

**E**

Eco-Labelling 39, 40, 42, 58
ecotoxicity 117
Electro-bioremediation 9, 15, 16
electrochemical oxidation 6
electrokinetic-*Fenton* 8, 15

Electrokinetic remediation vii, 1–8, 13–15, 17
Electro-osmosis 2, 3, 5, 13
Electrophoresis 3
environmental legislation 39
environmental management 41
Environmental remediation 108–111, 114, 116

**F**

Fenton's Reagent 75, 78, 82–85

**G**

green synthesis 112, 113
Groundwater v, vii

**H**

half-lives 62, 64, 68–70
heat cold storage 123, 128, 129
hydrogen peroxide 75, 78, 82, 84, 85, 89

**I**

*In Situ* Chemical Oxidation (ISCO) 75–78, 85, 87–90
*in situ* production 113
inversion of polarity 6
ion migration 2, 3

**L**

leak tests 39–41
leakage 41, 42, 44, 46
Life Cycle Assessment vi, vii, 95, 103, 104

**M**

Metabolite 63–66, 69

**N**

Nanomaterials 108–111, 114–117
Nanoremediation vi, vii, 108, 109, 115
natural extracts 112
nZVI synthesis 110

**O**

organic pollutants  75, 85–87
oxidant effectiveness  89, 90
oxidants  75, 76, 78–90
ozone  75, 78, 79, 81, 82, 86, 89, 90

**P**

permanganate  75, 78–81, 88–90
Petrol Stations  vii, 39–42, 44, 47, 51, 52, 56, 58
Petroleum hydrocarbons  139, 141, 142
Phytoremediation  144, 149, 151, 154–162

**R**

radius of influence  33, 34
remediation technologies  v, vi, 95, 104, 105
rhizoremediation  139, 144, 146

**S**

sodicity  149, 151–156
sodium adsorption ratio  150
Soil  i, iii, v, vii
soil salinity  149–154, 156
Soil Salinization  150, 160, 161
spillage  41, 47, 49
Surfactants  9, 10, 13, 14
Sustainability  95, 104
sustainable subsurface management  123

**U**

underground storage tanks  39, 40, 46

**Z**

zero-valent iron nanoparticles  108, 110, 114
zeta potential  5

Printed and bound by CPI Group (UK) Ltd, Croydon, CR0 4YY

01/11/2024

01782623-0016